8

Band 20 · August 1984

isotopenpraxis

AKADEMIE-VERLAG

ISSN 0021–1915 Isotopenpraxis, Berlin **20** (1984) 8, 285–320

EVP 9,– M

An unsere Autoren

Die Zeitschrift veröffentlicht Beiträge zu folgenden Themen: Kontrolle und Analyse radioaktiver und stabiler Isotope, Anwendung radioaktiver und stabiler Isotope in allen Gebieten der Forschung und Technik, der Industrie, Landwirtschaft und Medizin, Herstellung radioaktiver und stabiler Isotope sowie von markierten Verbindungen, kernelektronische Geräte für die Isotopenanwendung (Dickenmesser, Dichtemesser u. ä.) Detektoren für isotopentechnische Arbeiten und den Strahlenschutz in Isotopenlaboratorien, Meßtechnik und Strahlenschutz, Dosimetrie in Isotopenlaboratorien, kerntechnische Geräte für die Isotopenanwendung (Beton, Bestrahlungsanlagen u. ä.), Strahlenchemie, Radiochemie, ökonomische und perspektivische Fragen, Fragen der Weiterbildung.

1. Aufnahme von Manuskripten

Zur Veröffentlichung angenommen werden nur druckreife, noch nicht anderweitig publizierte Beiträge in deutscher, englischer oder russischer Sprache zum Themenbereich der Zeitschrift und zwar

Originalarbeiten
Kurze Mitteilungen
Aus der Praxis
Zusammenfassende Berichte
Tagungsberichte
Informationen.

Sind der Beitrag oder Teile daraus bereits veröffentlicht oder ist eine anderweitige Veröffentlichung vorgesehen, so muß dies der Redaktion mit genauer Angabe der Zweitveröffentlichung mitgeteilt werden. Mit der Annahme des Manuskriptes behält sich der Verlag das Recht der Vervielfältigung, Verbreitung und Übersetzung innerhalb der gesetzlichen Schutzfrist vor. Auszugsweiser Nachdruck oder photomechanische u. a. Wiedergabe sind an eine schriftliche Genehmigung des Verlages gebunden. Manuskripte, Zeichnungen und Photos von veröffentlichten Arbeiten werden nur auf den besonderen und bis zum Zeitpunkt der Veröffentlichung mitgeteilten Wunsch des Autors zurückgesandt.
Bei der unverlangten Einreichung von Manuskripten wird empfohlen, eine Befürwortung eines der Herausgeber oder Redaktionsausschußmitglieder beizufügen.
Manuskripte sind einzureichen bei der Redaktion, 1086 Berlin, Leipziger Str. 3–4.

2. Form der Manuskripte

Die Autoren werden gebeten:
2 Manuskripte, eines davon in Original-Maschinenschrift, einseitig und mit 2zeiligem Abstand einzureichen; gut leserliche Photokopien sind zulässig; auf dem 1. Blatt (unter Weglassung der Bezeichnung)
den Titel der Arbeit
Vorname-Initialen und Namen sämtlicher Autoren
gegebenenfalls die Institution, in der sie tätig sind
eine kurze Zusammenfassung der Arbeit von höchstens 15 Zeilen
Keywords in alphabetischer Reihenfolge und
als Fußnote die Anschrift der Institution anzugeben;
wenn möglich, auf gesonderten Blättern Namen, Titel der Arbeit und Zusammenfassung in den beiden anderen obengenannten Sprachen 2fach beizufügen;
Literaturhinweise (fortlaufend numeriert), Fußnoten, Tabellen, Abbildungen (in 2–3facher Veröffentlichungsgröße), Abbildungsunterschriften (getrennt von den Abbildungen mit Zeichenerklärungen, Quellenangaben, falls erforderlich, und gegebenenfalls mit einem Vermerk für besondere Wünsche bezüglich der Abbildungsgröße), private und Dienstanschrift der Verfasser mit akademischen Titeln und ausgeschriebenen Vornamen auf besonderen Blättern ans Ende des Manuskriptes zu stellen;

Wünsche für den Kolumnentitel (Seitenüberschrift) bekanntzugeben;
Absätze durch Einzüge von mindestens 3 Anschlägen kenntlich zu machen;
die Maßeinheiten des SI-Systems („Systeme International") zu verwenden;
Korrekturen im Manuskript unmittelbar an der zu korrigierenden Stelle anzubringen;

Formeln deutlich lesbar und möglichst groß einzutragen;
Farbabbildungen und nachträgliche Änderungen oder Ergänzungen in Abbildungen zu vermeiden;

hervorzuhebende Textstellen zu unterstreichen, und zwar halbfetten Satz mit einer geraden schwarzen Linie, kursiven Satz (Schrägschrift) mit einer blauen Wellenlinie (alle Eigennamen). Sperrungen gestrichelt;
besondere Buchstaben durch farbige Unterstreichung zu kennzeichnen, und zwar Formelzeichen in

Griechisch-Kursiv (α, β, ...) rot,
Kursiv (A, B, ...) blau
halbfetter Kursiv ($\boldsymbol{A}$, $\boldsymbol{B}$, ...) 2 Linien blau (Vektoren),
halbfetter Grotesk ($\mathbf{A}$, $\mathbf{B}$, ...) 2 Linien violett (Tensoren);

bei Literaturangaben auf Vollständigkeit zu achten. In nachstehender Reihenfolge müssen angegeben sein:

a) bei *Zeitschriften*: Vorname-Initialen und Namen sämtlicher Autoren, Titel der Zeitschrift (nach „Periodica Chimica" oder möglichst wenig gekürzt), Bandnummer (unterstrichen), Jahreszahl (in Klammern), Heftnummer (nur bei Zeitschriften, die jedes Heft mit S. 1 beginnen) sowie die Seitenzahl, auf der die zitierte Arbeit beginnt.

b) bei *Büchern*: Initialen der Vornamen und Name des Autors (Herausgeber werden durch den Zusatz „ed." gekennzeichnet), voll ausgeschriebener Titel des Buches in der Originalsprache, evtl. Bandnummer, voll ausgeschriebener Name des Verlages, Nummer der Auflage (außer der ersten), Erscheinungsort(e), Erscheinungsjahr (hier nicht in Klammern setzen!) und evtl. Kapitelnummer, Seitenzahl u. ä.

c) bei *Sammelbänden, Tagungsberichten*: Es soll zusätzlich der Titel des zitierten Beitrags angegeben werden. Die Seitenzahl muß unbedingt angegeben werden. Aus dem Zitat muß hervorgehen, ob es sich um einen Tagungssammelband oder um einen noch nicht veröffentlichten Vortrag handelt. Zwischen Tagungsort und Erscheinungsort des entsprechenden Tagungsberichts ist zu unterscheiden.

3. Korrekturen

Der Autor erhält zwei Umbruchabzüge, von denen einer möglichst schnell korrigiert an die Redaktion zurückgereicht werden soll. Auf Wunsch des Autors kann die Korrektur auch von der Redaktion gelesen werden.

Durch nachträgliche Manuskriptänderungen verursachte Korrekturkosten gehen, sofern sie 10% der Satzkosten übersteigen, zu Lasten des Autors.

Der Autor erhält 50 Sonderdrucke kostenlos. Weitere Sonderdrucke können bis zur Rückgabe der Korrekturabzüge auf Kosten des Autors bei der Redaktion bestellt werden

Bezugsmöglichkeiten

Bestellungen sind zu richten

- **in der DDR** an den Postzeitungsvertrieb unter Angabe der Kundennummer des Bestellenden oder an den AKADEMIE-VERLAG, DDR-1086 Berlin, Leipziger Straße 3–4, PF-Nr. 1233;
- **im sozialistischen Ausland** an eine Buchhandlung für fremdsprachige Literatur oder an den zuständigen Postzeitungsvertrieb;
- **in der BRD und Berlin(West)** an eine Buchhandlung oder an die Auslieferungsstelle KUNST UND WISSEN, Erich Bieber OHG, Wilhelmstraße 4–6, D-7000 Stuttgart 1;
- **in den übrigen westeuropäischen Ländern** an eine Buchhandlung oder an die Auslieferungsstelle KUNST UND WISSEN, Erich Bieber GmbH, Dufourstraße 51, CH-8008 Zürich;
- **im übrigen Ausland** an den Internationalen Buch- und Zeitschriftenhandel; den Buchexport, Volkseigener Außenhandelsbetrieb der Deutschen Demokratischen Republik, DDR-7010 Leipzig, Postfach 160; oder an den AKADEMIE-VERLAG, DDR-1086 Berlin, Leipziger Straße 3–4.

Zeitschrift „Isotopenpraxis"

Herausgeber: Im Auftrag des Akademie-Verlages herausgegeben von Prof. Dr. Justus Mühlenpfordt.
Verlag: Akademie-Verlag, DDR-1086 Berlin, Leipziger Straße 3–4, PF-Nr. 1233; Fernruf: 2236229; Telex-Nr.: 114420;
Bank: Staatsbank der DDR, Berlin, Kto.-Nr.: 6836-26-20712.
Verantwortlich für den Inhalt des Heftes: Prof. Dr. H. Hübner, Leipzig.
Anschrift der Redaktion (Fernruf Frau H. Schier: 6351177) der Zeitschrift Isotopenpraxis: Akademie-Verlag, DDR-1086 Berlin, Leipziger Str. 3–4, PF-Nr. 1233
Veröffentlicht unter der Lizenznummer 1527 des Presseamtes beim Vorsitzenden des Ministerrates der Deutschen Demokratischen Republik.
Gesamtherstellung: VEB Druckerei „Thomas Müntzer", DDR-5820 Bad Langensalza.
Erscheinungsweise: Die Zeitschrift „Isotopenpraxis" erscheint monatlich. Die 12 Hefte eines Jahrganges bilden einen Band. Bezugspreis je Band 216,– M zuzüglich Versandspesen Preis je Heft 18,– M.
Bestellnummer dieses Heftes: 1079/20/8.

AN (EDV) 12831

Isotopenpraxis 20 (1984) 8, S. 285 bis 289

Pituitary Gonodal Axis in Fertile and Infertile Human Males

A. A. Hafiez, K. Z. Mahmoud, E. Z. Abbas, F. Halawa

(Biochemistry and Andrology Departments, Fac. of Medicine, Cairo University, Cairo, Egypt)

Radioimmunoassay of serum level of PRL, LH, FSH, Testosterone and Estradiol were performed in normal fertile subjects and infertile patients (oligospermia and azospermia). The finding in normal fertile group may suggest that prolactin in human males has a role in steroidogenesis. Oligospermic and azospermic patients have hormonal pattern which is significantly higher than normal fertile group. Hyperprolactinemia was found in most cases of both infertile groups indicating that PRL has a significant role.

Bei normalen fruchtbaren und bei unfruchtbaren Versuchspersonen (Oligospermia und Azospermia) wurde ein Radioimmunoassay des Serumspiegels bezogen auf PRL, LH, FSH, Testosteron und Östradiol durchgeführt. Die bei der normalen fruchtbaren Personengruppe erzielten Ergebnisse könnten darauf hindeuten, daß das Prolactin bei Männern hinsichtlich der Steroidgenese eine Rolle spielt. Oligospermische und azospermische Patienten haben einen Hormonspiegel, der wesentlich höher liegt als bei der normalen fruchtbaren Gruppe. Dabei ist die Abweichung vom Normalwert bei Oligospermia höher und bei Azospermia wesentlich höher.

Keywords

gonads; hypothalamus; lth; pituitary glands; radioimmunoassay; steroid hormones

Introduction

The reproductive axis in men includes the extrahypothalamic central nervous system, hypothalamus, pituitary gland and testis. Hypothalamus stimulates the pituitary by releasing factors that secrete gonadotropins (LH and FSH), which regulate the endocrine and gametogenic functions of mammalian testis. LH stimulates production of testosterone by Leydig cells which is essential hormone for spermatogenesis in human and rodents [1]. FSH may be involved in the synthesis, transport and metabolism of androgens in the testis [2].

Prolactin was proved to be present in the serum of males and its level is comparable to or somewhat lower than those observed un non-pregnant non-lactating females [3]. The role of prolactin in males was first suggested by the fact that treatment of dwarf male mice with purified prolactin increased the weight of testis, accessory reproductive glands and increased the number of spermatocytes and spermatids and induced fertility [4]. This was confirmed in hypophysectomised rate and it was suggested that prolactin potentiates the action of LH on testicular steroidogenesis [5].

These informations stimulated our interest to study the serum hormonal levels of prolactin, LH, FSH, testosterone and estradiol in normal fertile human subjects and in fertile oligospermic and azospermic patients. This study was also conducted to investigate the possible role of prolactin in regulation of the human testis and to elucidate its role in production of oligospermia and azospermia.

Material and Methods

Hundred and ten adult men were the subjects of the present study who were classified as follows:

Normal fertile control comprising 30 healthy fertile men, ageing 21 — 37 years and fathering children at least one child not more than one year old.

Oligospermic group comprising 42 cases of the same age group with reduced sperm count (below 20 millions/ml).

Azospermic group comprising 38 cases of the same age group.

Semen was analysed fir full andrological picture dor all cases after 4 — 5 days abstinence. Blood samples were obtained at 10·0 o'clock a.m. and allowed to clot and serum was separated and kept frozen at -20 °C for radioimmunoassay of LH, SFH, prolactin, testosterone and estradiol by CIS kits.

The data obtained were analysed by student "t" test for significance of the difference. Correlation between different parameters was also calculated.

Results

The data are presented in Tables 1 — 3 and Figures 1 — 9.

The serum hormonal levels in normal fertile Egyptian men are shown in Tab. 1. The mean values (ng/ml) were $3{\cdot}348 \pm 1{\cdot}64$ (S.D.) for prolactin (PRL), $2{\cdot}606 \pm 1{\cdot}134$ (S.D.) for LH, $1{\cdot}623 \pm 0{\cdot}916$ (S.D.) for FSH, $4{\cdot}60 \pm 2{\cdot}135$ (S.D.) for testosterone (T) and $0{\cdot}0295 \pm 0{\cdot}006$ (S.D.) for estradiol (E_2).

The results show that both infertile groups (oligospermia and azospermia) have a characteristic hormonal patterns which are essentially higher than that of normal fertile controls (Tab. 1 and Fig. 1). With the exception of LH, all other serum hormonal levels showed a significant increase in oligospermic patients. In group of azospermia there is a significant increase in the serum levels of PRL, LH and FSH compared to normal and in LH and FSH compared to oligospermic group. The serum levels of testosterone and estradiol in azospermic patients is not changed significantly from the normal values but show a significant drop compared to that of oligospermia.

The ratios of the different serum hormonal levels relative to testosterone were calculated. In the normal fertile group it is $1{\cdot}03 \pm 0{\cdot}47$ for RRL/T, $0{\cdot}87 \pm 0{\cdot}72$ for LH/T, $0{\cdot}58 \pm 0{\cdot}59$ for FSH/T and $0{\cdot}012 \pm 0{\cdot}007$ for E_2/T. In the oligospermic group there is a significant drop in PRL/T

Tab. 1. Comparative Data for Normal Subjects Oligospermic and Azospermic cases

	Volume	Count mil/ml	Spermiogram Motolity	Viability	Abnormal form	Pus cells	Normal level ng/ml PRL	LH	FSR	T	E_2
Mean	3.51	69.5	71.3	33.33	9.13	3.3	3.348	2.606	1.623	4.360	0.295
Normal ± SDM	1.46	12.7	7.6	4.33	3.59	1.5	1.64	1.134	0.916	2.135	0.008
(30) ± SEM	0.27	2.3	1.4	0.83	0.66	0.2	0.28	0.206	0.169	0.489	0.002
Mean	2.5	6.10	33.20	13.8	23.10	11.40	4.302	3.146	2.703	6.266	0.058
Oligospermic ± SDM	1.1	6.30	20.10	10.8	14.40	11.75	2.69	2.175	1.810	3.288	0.021
Cases ±SEM	0.17	0.87	0.32	1.57	2.22	1.81	0.42	0.320	0.290	0.510	0.004
(42) p value	<0.001	0.001	0.001	0.001	0.001	<0.001	0.025	0.20	0.005	0.001	0.001
Mean	1.37	—	—	—	—	9.63	4.688	5.535	4.287	3.126	0.025
Azospermic ± SDM	4.43	—	—	—	—	20.5	2.800	3.769	3.279	2.746	0.009
Cases											
(38) ± SEM	0.719	—	—	—	—	3.33	0.455	0.611	0.532	0.445	0.001
p	0.001	—	—	—	—	0.001	0.02	0.001	0.001	0.2	0.5
P_1	0.1	—	—	—	—	0.7	0.6	0.001	0.001	0.02	0.001

PRL = Prolacting; LH = Luteining Hormone; FSR = Follicle Stimulating Hormone
T = Testosterone; E_2 = Estradiol
p = as compared to normal subjects
P_1 = as compared to oligospermic cases

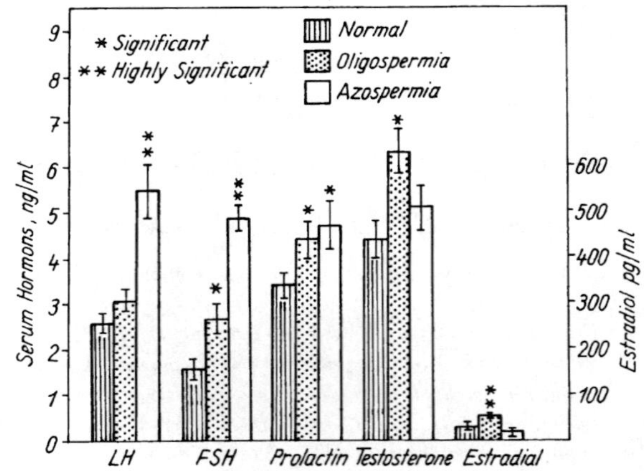

Fig. 1. Showing the mean values of serum hormonal changes in normal subjects (30) oligospermic (42) and Azospermic patients (38)

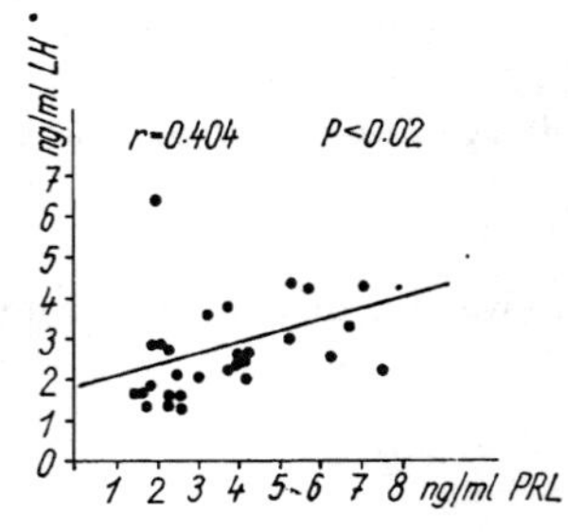

Fig. 2

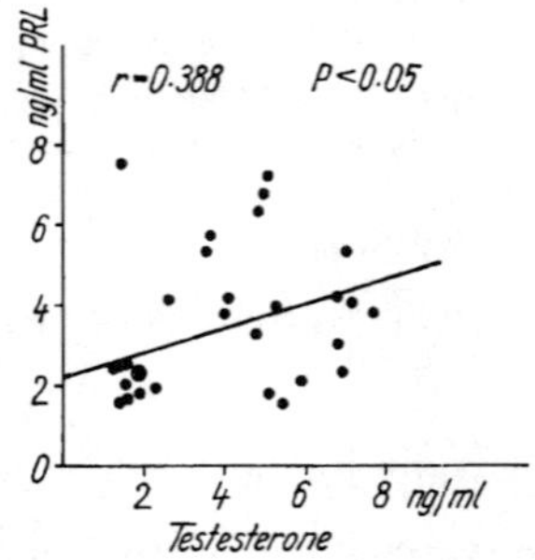

Fig. 3

Fig. 2. Correlation between PRL and LH in normal fertile men
Fig. 3. Correlation between testosterone and PRL in normal fertile men

ratio (0.76 ± 0.41, $P < 0.025$) and in LH/T (0.52 ± 0.38, $P < 0.01$), with no changes in FSH/T and E_2/T ratio. In the azospermic group, we have found a significant decrease in PRL/T ratio (0.07 ± 0.30, $P < 0.005$) and E_2/T ratio (0.006 ± 0.004, $P < 0.01$)), but without significant changes in LH/T and FSH/T when compared to normal. By comparing the azospermic group to that of oligospermia, we detected a significant increase in LH/T ratio ($P < 0.005$) and FSH/T ($P < 0.005$) with decrease in E_2/T ratio ($P < 0.01$).

The correlation between the serum levels of different hormones in each group were studied (Tab. 1). In normal fertile group a positive correlation was found between PRL and LH ($r = 0.404$, $P < 0.02$, Fig. 2) as well as between T and PRL ($r = 0.388$, $P < 0.05$, Fig. 3) and a negative correlation between E_2 and LH ($r = 0.265$, $P < 0.05$, Fig. 4).

In oligospermic group: there were a positive correlation between T and FSH ($r = 0.366$, $P < 0.05$, Fig. 5), E_2 and

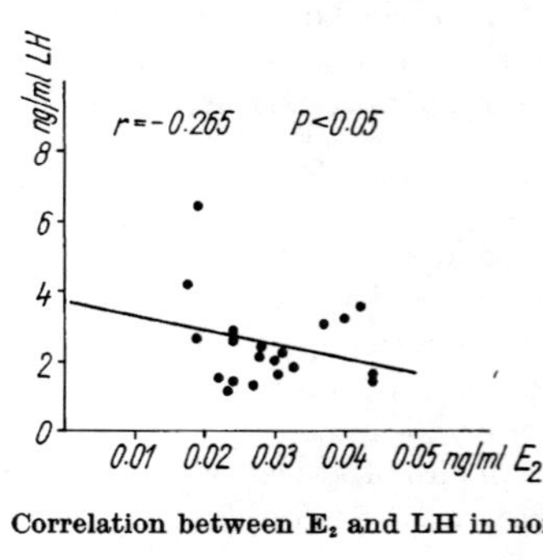

Fig. 4. Correlation between E_2 and LH in normal fertile men

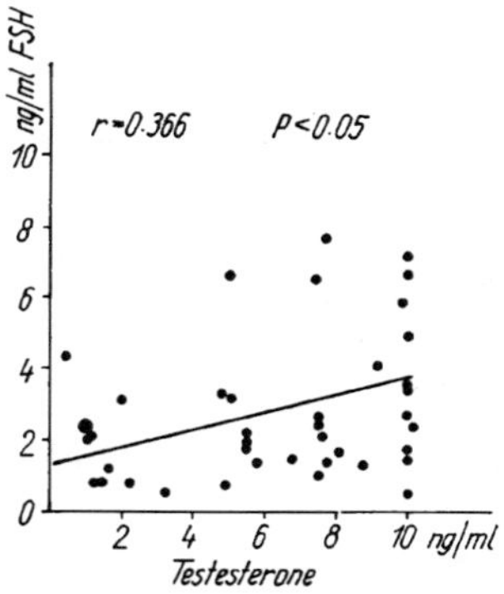

Fig. 5

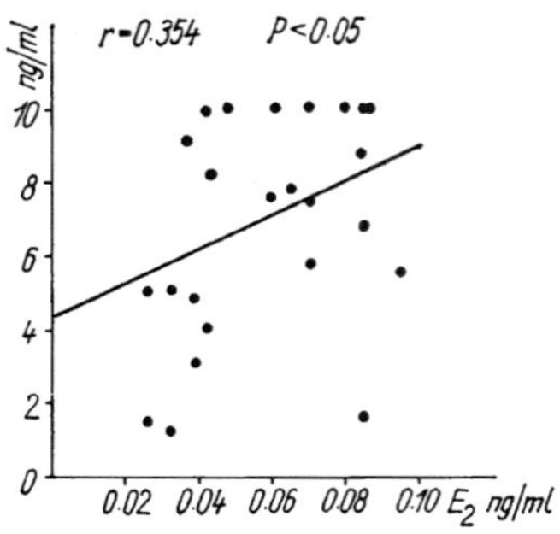

Fig. 6

Fig. 5. Correlation between testosterone and FSH in oligospermic cases
Fig. 6. Correlation between E_2 and testosterone in oligospermic cases

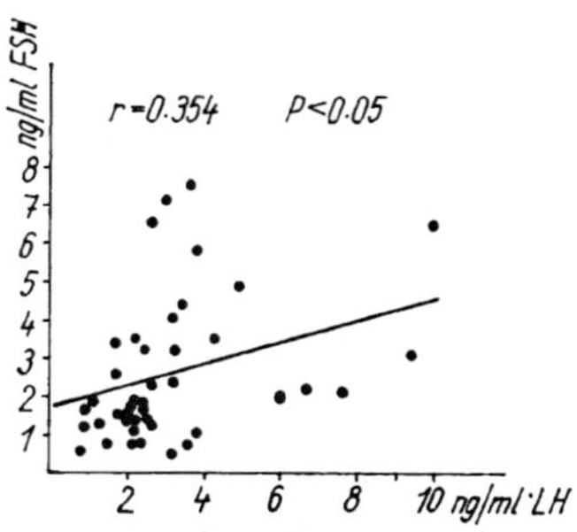

Fig. 7. Correlation coefficient between LH and FSH in oligospermic cases

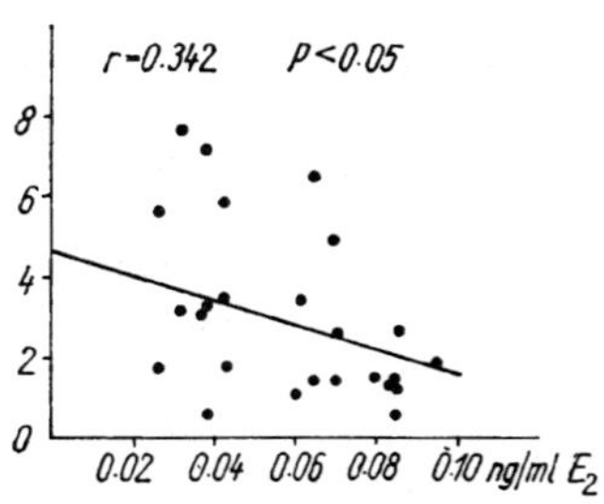

Fig. 8. Correlation between E_2 and FSH in oligospermic cases

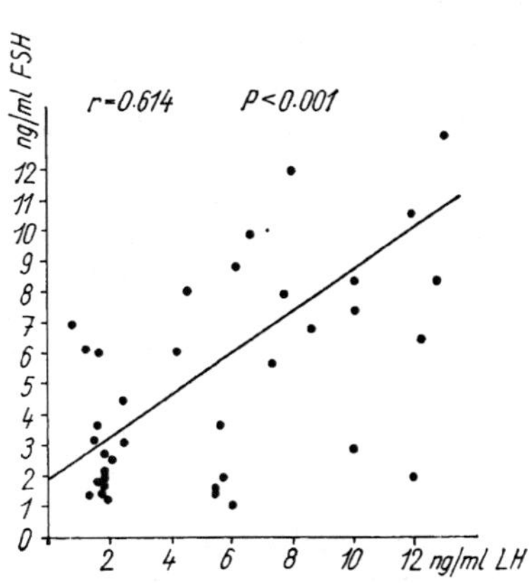

Fig. 9. Correlation coefficient between LH and FSH in azospermic cases

T ($r = 0.354$, $P < 0.05$, Fig. 6), LH and FSH ($r = 0.354$, $P < 0.05$, Fig. 7) and E_2 and FSH ($r = 0.342$, $P < 0.05$, Fig. 8).

In azospermic group: there is only a positive correlation between LH and FSH ($r = 0.614$, $P < 0.001$, Fig. 9).

Discussion

Our data for semen analysis in normals pointed out that sperm viability compatable with fertility potentials should start with 55% motile sperms after one hour and at least 25% motile sperms after 3 hours (Tab. 1). These data coinside with other reports [6]. The sperm density varied between 45–90 million/ml, the determining point to segregate fertile from non-fertile by sperm density was found to be 40 million/ml which agrees with the report of *Macleod* [6]. Recently, it was claimed that 25 million/ejeculate can be considered the lowest level for normal [7]. However, *Smith* et al. [8] consider sperm viability rather than sperm density to be important for fertility potentials. We believe that fertility potential seems to be multifactorial comprising sperm viability, fertilizing capacity and sperm density.

Tab. 2. Changes in hormonal ratios in normals and infertile cases

	Prolactin	LH	FSH	Estra-diol
	testo.	testo.	testo.	testo.
Normal				
M	1·030	0·87	0·58	0·012
SDM	0·470	0·72	0·59	0·007
SEM	0·085	0·13	0·11	0·002
Oligo-spermia				
M	0·76	0·52	0·50	0·011
SDM	0·41	0·38	0·42	0·009
SEM	0·08	0·06	0·07	0·002
P	<0·025	<0·01	<0·05	<0·7
Azo-spermia				
M	0·70	0·86	0·71	0·006
SDM	0·30	0·50	0·46	0·004
SEM	0·05	0·09	0·09	0·001
P_1	<0·005	<0·95	0·3	<0·01
P_2	<0·5	<0·005	<0·05	<0·01

testo. = Testosterone
P_1 = Significance with normal
P_2 = Significance with oligospermia

The normohormongram (Tab. 3) showed a positive significant correlation between serum prolactin and LH levels (Fig. 2) as well as between prolactin and testosterone (Fig. 3). These findings in normal fertile group may suggest that prolactin in human males has a role in potentiating the action of LH on steroidogenesis in human testis. A similar effect was described for prolactin and LH on testis of rats [5]. This suggestion was confirmed by *Yen* and *Jeffe* [9] who stated that the main site of action of LH is the Leydig cell enhancing steroidogenesis. It is also reported that radioiodinated prolactin receptors, shown by autoradiography, to be localized around the *Leydig* cells of the mouse and rat testis [10].

Serum hormonal levels in oligospermic cases were found to be elevated significantly as compared to normal level (Tab. 1 and Fig. 1). Similar results was also reported by many investigators [11–13].

Tab. 3. Corrleation coefficient between different hormone levels in the different groups investigated

Hormones	Normal		Oligospermia		Azospermia	
	r	*P* value	*r*	*P* value	*r*	*P* value
1 — Testosterone and prolactin	0·388	0·05*)	0·117	0·1	0·023	0·1
2 — Testosterone and LH	0·067	0·1	0·11	0·1	0·032	0·1
3 — Testosterone and FSH	0·117	0·1	0·366	0·05*)	0·121	0·1
4 — Testosterone and Estradiol	0·186	0·1	0·354	0·05*)	0·025	0·1
5 — Prolactin and LH	0·404	0·02*)	0·179	0·1	0·02	0·1
6 — Prolactin and FSH	0·093	0·1	0·013	0·1	0·067	0·1
7 — Prolactin and Estradiol	0·030	0·1	0·020	0·1	−0·043	0·1
8 — LH and FSH	0·0183	0·1	0·354	0·05*)	0·614	0·001*)
9 — LH and and Estradiol	−0·265	0·05*)	0·168	0·1	−0·099	0·1
10 — FSH and	0·027	0·1	0·342	0·05*)	−0·075	0·1

*) significant correlation

A similar rise was detected in the serum hormonal levels in azospermic group with the exception of estradiol which was found to be lower than normals.

By calculating the ratio of different hormonal levels to testosterone (Tab. 2), we found that prolactin/testosterone and LH/testosterone were significantly lower in oligospermic cases when compared to normals. No significant changes in FSH/testosterone and E_2 testosterone were found.

A positive significant correlation was found between FSH and testosterone (Fig. 5) and between testosterone and E_2 (Fig. 6) in oligospermic which indicate the proportional changes in these hormones, so no changes could be detected in FSH/testosterone and E_2/testosterone ratios.

Those findings point to the normal functions of *Leydig* cells (Steroidegenesis) in response to prolactin and LH as evidenced by a higher production of testosterone and E_2.

The significant drop of serum levels of testosterone and estradiol in azospermic group relative to oligospermic group indicates that the steroidogenic function of *Leydig* cells are affected. This is also reflected by the increase in LH/testosterone and FSH/testosterone ratios.

Hyperprolactinemia has been recently reported in different cases of infertility [13]. Hyperprolactinemia in men can be associated with hypogonadism and impotence [14]. Prolactin seems to alter androgen metabolism of the male accessory sex glands [15]. In contrast, chronic hyperprolactinemia induced in rats and mice by multiple ectopic pituitary isografts does not affected plasma testosterone level or the weight of the testis [16]. This work provide an evidence in favour of destructive effects of hyperprolactinemia as it showed that increased prolactin levels induces variable effects on spermatogenesis in men. Prolactin by itself might not be the only cause for these changes as it is well known that regulation of testicular function is affected by a complex of the pituitary hormones rather than by a single hormone. The increase in LH and FSH levels concurrent with hyperprolactinemia support this statement.

The increase in plasma LH, FSH and prolactin with increase in testosterone and E_2 levels in the infertile group investigated is in favour of disturbed feed back mechanism. It is well known that testicular steroids mediate feed back inhibition of pituitary gonadotropins secretion in the male. The presence of both androgen and estrogen receptors in the hypothalamus and pituitary support this view [17]. This is further confirmed in men [18].

This shows that in spite of normal functioning pituitary and *Leydig* cells, there is failure of seminiferous tubule function. The failing tubules might not be capable of producing inhibitory internal factor that regulate the pituitary hormones (inhibin theory). It seems that the pituitary hormones are modulated by tubular function and to a certain extent by testosterone secretion. This agrees with the work of *Lugaro* et al. [19] who claimed that administration of inhibin induce 50—80% inhibition of FSH release in rats. Inhibin action might be through its effect on the hypothalamic releasing factors and thus it affect the hypothalamic-gonadal axis.

We raise the suggestion that the pituitary-gonadal functions is regulated in the way that pituitary gonadotropins (LH and FSH) and prolactin stimulate the *Leydig* cells to secrete testosterone. Testosterone or its active metabolites stimulate the tubular cells that respond by releasing inhibitory factors that cause inhibition of hypothalamic-pituitary functions. So, failure of synthesis or secretion of this inhibitory factors will result in an increased level of FSH, LH, PRL and testosterone.

In our cases the failure of tubular functions may be due to defect in the binding sites of androgenic receptors or failure to metabolise testosterone in the normal way.

Thus, we can conclude that infertility in our oligospermic and azospermic cases may be induced by a cause (s) at the gonadal level but not at the hypothalamic hypophyseal levels since the gonadotropins as well as prolactin and testosterone are all elevated.

Received June 29, 1983
Accepted in revised form December 20, 1983

References

[1] *M. L. Dufau, K. Watanable, K. J. Catt,* Endocr. **92** (1972) 6
[2] *B. I. Posner, P. A. Kelly, R. P. C. Shir, H. O. Friesen,* Endo. **95** (1974) 421
[3] *A. G. Frantz, D. L. Kleinberg, G. E. Noel,* Recent Prog. Horm. Res. **28** (1982) 527
[4] *A. Bartke, C. W. Lioyd,* J. Endocr. **46** (1970) 313
[5] *A. A. Hafiez, A. Bartke, C. W. Lioyd,* Brit. J. End. **52** (1972) 327
[6] *J. Maclead,* Fertil. and Steril. Jerusalem, Israel, Academy of Science (1970) 268
[7] *Z. Zukerman, L. J. Rodriquez-Rigan, E. Steinberger,* Fertil and Sgeril. **28** (1977) 1310
[8] *K. D. Smith, R. K. Tcholakian, M. Chowdhury, F. Steinberger,* Fertil. and Steril. **25** (1974) 965
[9] *S. S. C. Yen, R. B. Jaffe,* Reprod. Endocr. W. B. Saunders Company, Philadelphia/London/Toronto 1978
[10] *H. Ragamemi, A. Oksanen, T. Vanha-Pertlula,* Horm. Res. **5** (1974) 6
[11] *S. Segal, W. Z. Polishuk, M. Ben-Daris,* Fertil. and Steril. **27** (1976) 1425
[12] *W. Krause,* Fortschr. Med. **95** (1977) 1984
[13] *K. Saidi, B. Wenn, F. Sharif,* Lancet **99** (1977) 250
[14] *R. M. Boyar, S. Kapen, M. Perlow, J. F. Sassin, L. Hellman,* J. Clin. Invest. **53** (1974) 1588
[15] *D. Helmrich, M. Altwein,* Urol. Res. **4** (1976) 101
[16] *A. Batrke, M. S. Smith, P. G. Peron, S. Dalterio,* Endocr. **100** (1977) 1982
[17] *B. Attardi, L. N. Geller, S. Ohno,* Endocr. **98** (1978) 864
[18] *S. P. Marynick, D. L. Loriauk, M. B. Lipsett,* J. Clin. Endocr. Metab. **39** (1979) 396
[19] *G. Lugaro, G. Fachini, L. Riboni, M. A. Bacigalupo,* J. Reprod. Fertil. Suppl. **26** (1979) 193

Isotopenpraxis **20** (1984) 8, S. 289 bis 292

The Effect of Control of Diabetes Mellitus on Plasma T_4, T_3, rT_3 Levels and Half Muscle Relaxation Period

A. A. Ismail, A. A. Hafiez, S. N. Sayed, E. Z. Abbas, F. A. Halawa, M. M. Youssef

(Biochemistry and Internal Medicine Departments, Faculty of Medicine, Cairo University, Cairo, Egypt)

Twenty five diabetic patients of the maturity onset type who showed no clinical evidence of either peripheral neruropathy or diabetic amyotrophy were selected for this study. All patients were subjected to the following investigations: estimation of half muscle relaxation period of the quadriceps muscle knee jerk, measurement of plasma levels of thyroxine (T_4), triiodothyronine (T_3) and reverse triiodothronine (rT_3), determination of fasting and two hours postprandial blood sugar levels. The quadriceps mucle relaxation period in uncontrolled diabetics was significantly longer than in normals. Control of diabetes by glibenclamide or gliclazide did not cause a significant change in muscle relaxation period. There was also no significant difference between the effects of the two drugs.

Für diese Untersuchung wurden 25 Diabetiker am Reifebeginn, die keinerlei Anzeichen von peripherer Neuropathie oder diabetischer Amyotrophie zeigten, ausgewählt. Alle Patienten wurden folgenden Untersuchungen unterzogen: Schätzung der halben Entspannungszeit beim Kniemuskel, Messung der Niveaus von Thyroxin (T_4), Trijodthyronin (T_3) und Umkehr-Trijodthronin (rT_3) im Plasma, Ermittlung des Blutzuckerspiegels nach dem Fasten und zwei Stunden nach der Nahrungsaufnahme. Die Kniemuskelentspannungszeit war bei unbehandelten Diabetikern wesentlich länger als bei Normalpersonen. Eine Behandlung der Diabetes mit Glibenclamid oder Gliclazid führte zu keiner wesentlichen Veränderung der Muskelentspannungszeit. Es bestand auch kein wesentlicher Unterschied zwischen beiden Drogen hinsichtlich ihrer Wirkung.

Keywords

diabetes mellitus; patients; thyroid hormones; thryoxine; triiodothyronine; TSH

Introduction

In a previous work, the thyroid functions were studied in a group of insulin-dependent young diabetics before and after control of their diabetes by insulin [1]. The data obtained in their study showed a lowered level of T_3 and T_4 and a raised level of TSH in uncontrolled young diabetics, and that control of diabetes by insulin resulted in an increase in T_3 and T_4 levels, with concomitant decrease in the TSH levels to reach nearly normal values.

The object of this work is to find the effect of short period control of the diabetic condition by oral hypoglycemic drugs, namely glibenclamide or gliclazide, on thyroid function.

Tendo-Achilles reflex has been originally invented to detect the thyroid activity. It was found that some clinically euthyroid diabetics have prolonged muscle relaxation [2–4]. This finding was confirmed by a number of observers who have related it to the diabetic condition itself [2].

In the last few years, it was discovered that some diabetics have difficulty in the deiodination of T_4 to the active metabolite T_3 [1, 5]. In view of this knowledge, the possibility that tendon jerk relaxation defect in diabetics is due to masked subtle hypothyroidism rather than diabetes itself came into light. Accordingly, we studied the effect of short period control of diabetes, by oral hypoglycemic drugs, on the muscle relaxation period and plasma levels of T_3, T_4 and rT_3.

Materials and Methods

Twenty five diabetic patients of the maturity onset type, aged between 35–55 years have been selected. Twenty patients (80%) disclused their diabetic condition within the past few days before the study and thus had never received oral hypoglycemic drugs. Five patients (20%) had their diabetic condition discovered within the last few months (3–6 months) and have had short and irregular courses of oral hypoglycemic drugs. None of these patients showed clinical evidence of peripheral neuropathy nor of diabetic amyotrophy. The patients were classified into two groups:

The first group included 13 patients who received glibenclamide and the second group comprised 12 patients who received gliclazide each for two weeks. All of them were subjected to the following tests before and two weeks after control of diabetes by the drugs:

1. Estimation of the fasting and two hours postprandial blood glucose levels by glucose oxidase (using kits produced by Boheringer Mannheim).
2. Determination of plasma levels of T_4, T_3 and rT_3 by radioimmunoassay (using kits produced by Serono-Biodata Diagnostics, Italy).
3. Determination of the half relaxation period of the quadriceps muscle (knee jerk) by photoelectric device of *Gilson* [6] using the modification made by *Habib* [4] and *El-Badawy* [3].

Ten normal individuals were selected as a control group and were subjected to the same investigations.

The results are as seen in Tab. 1—4 and Fig. 1—3.

Analysis of Results and Discussion

Hafiez et al. [1] found that T_4 production by the thyroid gland and possible T_4 to T_3 conversion are both impaired in uncontrolled insulin dependent young diabetic patients and that this impairment could be corrected by insulin therapy.

In this work we studied the effect of control of diabetes by oral hypoglycemic drugs on the plasma levels of T_4 T_3 and rT_3 and on the quadriceps half muscle relaxation periods.

Our results show that the mean plasma levels of T_4 and T_3 in uncontrolled diabetics are not significantly different from the mean plasma levels in normal subjects. The mean plasma level of rT_3 is however, significantly increased in uncontrolled cases when compared to normals. Inspite of the normal plasma level of thyroxine, the quadriceps half relaxation time in diabetics before control of the diabetic state is significantly longer than in normals (Tab. 1, 2 and Fig. 1). The later finding agrees with that previously reported by *Habib* [4], *El-Badawy* [3] and reports of *Beardwood* and *Schumbacher* [2] on the ankle jerk. However, it is not in confirmity with the work reported by *Hunton* [7] who found the delay in relaxation is limited to those receiving oral antidiabetic drugs as none of our cases

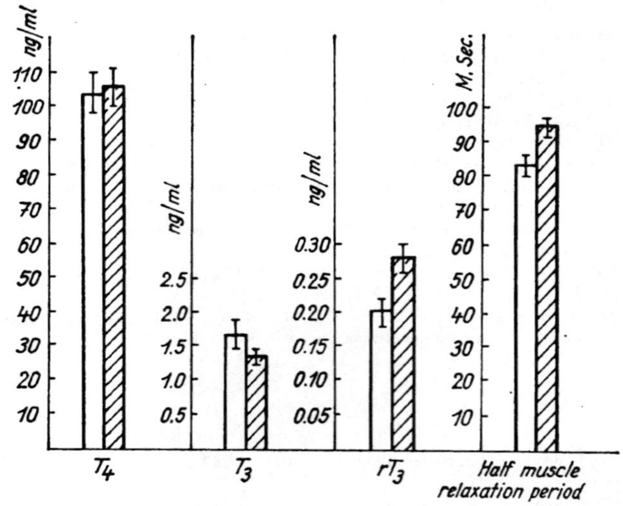

Fig. 1. Mean values (±SE) of plasma T_4, T_3 and rT_3 and half muscle relaxation period in normals and diabetics (hatched columns)

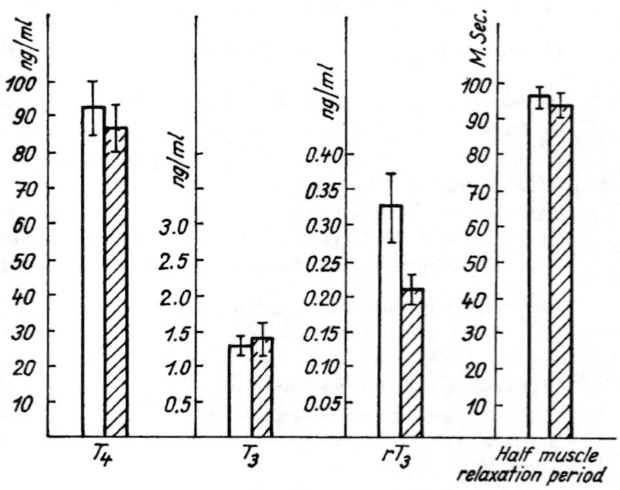

Fig. 2. Mean values (±SE) of plasma T_4, T_3 and rT_3 and half muscle relaxation period in diabetics before and after (hatched columns) Glibenclamide

Tab. 1. Plasma T_4, T_3 and rT_3 levels and half muscle relaxation period in normal cases

Cases No.	Plasma T_4 ng/ml	Plasma T_3 ng/ml	Plasma rT_3 ng/ml	1/2 muscle relaxation period "m · sec"
1	140	2·30	0·21	90
2	86	2·50	0·18	80
3	79	2·10	0·1	90
4	110	0·09	0·18	80
5	102	1·70	0·24	90
6	100	1·75	0·3	80
7	67	0·90	0·25	70
8	85	2·00	0·15	80
9	80	0·95	0·3	80
10	87	1·50	0·1	90
Mean	93·6	1·66	0·201	83·0
S.D. ±	20·5869	0·5878	0·0726	6·749
S.E. ±	6·51	0·1862	0·023	2·1344

Tab. 2. Plasma T_4, T_3, rT_3 levels ad the half-muscle relaxation period in uncontrolled diabetics

Case No.	Plasma T_4 ng/ml	Plasma T_3 ng/ml	Plasma rT_3 ng/ml	$^1/_2$ muscle relaxation period "m · s"
1	59	2·00	0·23	90
2	118		0·24	80
3	74	1·25	0·17	80
4	140	0·60	0·27	80
5	99	1·00	0·22	110
6	94	0·60	0·31	110
7	160	2·50	0·4	100
8	101	1·25	0·22	80
9	79	2·40	0·14	90
10	69	1·45	0·15	80
11	84	1·00	0·25	110
12	115	1·20	0·24	110
13	66	0·90	0·9	100
14	90	1·00	0·16	80
15	117	1·20	0·3	100
16	145		0·36	90
17	64		0·15	100
18	95	1·10	0·23	100
19	92	1·80	0·24	100
20	75	1·40	0·25	100
21	150	0·90	0·4	100
22	70	1·75	0·4	90
23	74	2·00	0·32	80
24	80	1·70	0·28	
25	84	0·95	0·24	
Mean	95·76	1·36136	0·2828	94·35
S.D. ≦	28·624	0·5322	0·1495	11·6
S.E. ≦	5·7249	0·1134	0·0289	2·419
p°	>0·35	>0·05	<0·025	<0·0025

P° = Uncontrolled diabetics Vs normal subjects.

Tab. 3. Plasma T_4, T_3 and rT_3 levels and half-muscle relaxation period, before and after Glibenclamide

Case No.	Plasma T_4 ng/ml	ng/ml	Plasma T_3 ng/ml		Plasma rT_3 ng/ml		$^1/_2$ muscle relaxation period, "m · s"	
	before	after	before	after	before	after	before	after
1	66	103	0·90	1·00	0·9	0·32	100	80
2	90	61	1·00	2·20	0.16	0·14	80	100
3	117	108	1·20	1·30	0·3	0·2	100	100
4	145	103			0·36	0·23	90	90
5	64	110			0·15	0·17	100	80
6	95	85	1·10	1·90	0·23	0·14	110	110
7	92	58	1·80	2·10	0·24	0·15	100	100
8	75	90	1·40	2·50	0·25	0·12	100	100
9	150	125	0·90	0·50	0·4	0·3	90	90
10	70	100	1·75	0·70	0·4	0·32		
11	74	53	2·00	1·00	0·32	0·24	80	80
12	80	85	1·70	2·00	0·28	0·2		
13	84	61	0·95	0·50	0·24	0·22	100	100
Mean	92·46	87·85	1·336	1·427	0·325	0·212	95·46	93·64
S.D.	28·25	23·19	0·4099	0·7336	0·29	0·069	9·34	10·26
S.E.	7·834	6·431	0·1236	0·2211	0·053	0·019	2·59	2·845
P	>0·1		>0·35		<0·05		>0·3	

Dose: Enough dose was given to control the individual diabetic.

Tab. 4. Plazma T_4, T_3 and rT_3 levels and half-muscle relaxation period before and after Gliclazide

Case No.	Plasma T_4 ng/ml		Plasma T_3 ng/ml		Plasma rT_3 ng/ml		$^1/_2$ muscle relaxation period, "m · s"	
	before	after	before	after	before	after	before	after
1	59	58	2·00	1·20	0·23	0·2	90	100
2	118	59			0·24	0·1	80	90
3	74	67	1·25	2·00	0·17	0·18	80	80
4	140	102	0·60	0·55	0·27	0·2	80	110
5	99	93	1·00	2·50	0·22	0·22	110	100
6	94	66	0·60	0·95	0·31	0·09	110	100
7	160	170	2·50	0·65	0·4	0·27	100	100
8	101	79	1·25	0·90	0·22	0·13	80	80
9	79	68	2·40	2·00	0·14	0·17	90	90
10	69	68	1·45	1·30	0·15	0·13	80	80
11	84	93	1·00	1·05	0·25	0·2	110	90
12	115	94	1·20	2·25	0·24	0·21	110	80
Mean	99·33	84·75	1·386	1·395	0·237	0·175	93·43	91·667
S.D.	29·84	30·79	0·6523	0·6754	0·07	0·063	13·7	10·39
S.E.	8·61	8·8	0·1957	0·2036	0·02	0·015	3·95	2·973
P		>0·1		>0·47		<0·025		>0·35

Dose: Enough dose was given to control the individual diabetic.

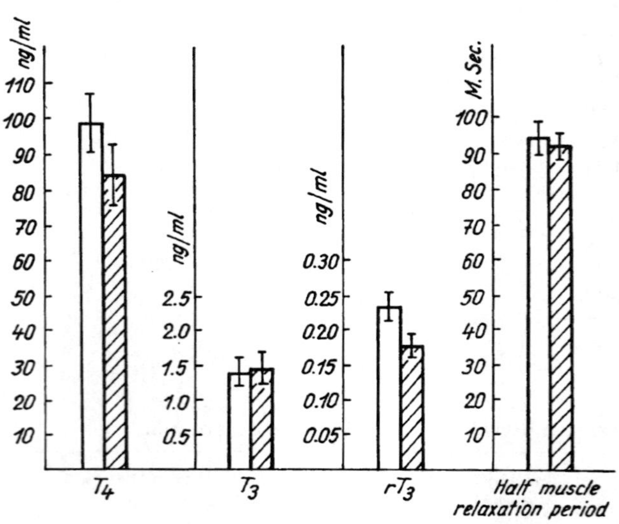

Fig. 3. Mean values (±SE) of plasma T_4, T_3 and rT_3 and half muscle relaxation period in diabetics before and after (hatched columns) gliclazide

had oral antidiabetics for considerable duration before the experiment.

Control of the diabetes by either glibenclamide (Tab. 3 and Fig. 2) or gliclazide (Tab. 4 and Fig. 3) resulted in a significant drop of rT_3 level in plasma, while it had no effect on either plasma levels of T_4 and T_3 or the muscle relaxation period.

This shows that normalization of carbohydrate tolerance is accompanied with reduction in rT_3 in plasma. It is known that rT_3 increases during starvation, carbohydrate deficiency and in cases of ketosis [5, 8]. As in diabetic patients one can consider the case as a sort of "carbohydrate cell starvation", release of insulin by oral hypoglycemic dugs may control this condition reducing the rT_3 level.

The lack of correlation between the muscle relaxation period and the control of diabetes agrees with reports given by *Beardwood* and *Schumbacher* [2]. They suggested that the delay in muscle relaxation in diabetics may be due to defect in the intracellular metabolism and that longer periods of therapy are required for correction of these abnormal metabolic pathyways. *Reinfrank* et al. [9] reported three patients with newly discovered diabetes showing return of abnormally prolonged half muscle relaxation periods to the predicted normal range values after four months of treatment with diet and tolbutamide.

In conclusion, the work shows that, in uncontrolled diabetics of the maturity onset type there is an increase in the plasma levels of rT_3 and prolongation of half muscle relaxation period. Also, this study shows that control of diabetes by oral hypoglycemic drugs, namely glibenclamide or gliclazide, for a short time resulted in drop of the elevated rT_3 levels but had no effect on the half muscle relaxation period and no significant difference was detected between the effect of the two drugs used in this study.

Received March 16, 1983
Accepted in revised form November 2, 1983

References

[1] *A. A. Hafiez, A. A. Ismail, M. R. Awadeen, E. Z. Abbas, M. S. A. Farag,* The third scientific conference of the Egyptian Society of Basic Medical Sciences, 1981

[2] *D. H. Beardwood, L. H. Schumacher,* Amer. J. Med. Sci. **247** (1964) 324

[3] *S. A. El-Badawy,* Faculty of Medicine, Cairo University, M. D. Thesis, 1973

[4] *M. Habib,* Faculty of Medicine, Cairo University, M. D. Thesis, 1973

[4] *C. S. Pittman, A. K. Suda, J. B. Chambers, H. G. MeDaniel, G. Y. Rey, B. K. Preston,* J. Clin. Endocrin. **48** (1979) 854

[6] *W. E. Gilson,* New Eng. J. Med. **260** (1959) 10257

[7] *R. B. Hunten, M. V. Wells, E. W. Skripper,* Lancet **2** (1965) 449

[8] *S. H. Ingbar, N. Freinkle,* Recent Proger. Res. **16** (1969) 353

[9] *R. F. Reinfrank, R. P. Kaufman, H. J. Wetsone, J. A. Clennon,* J.A.M.A. **199** (1967) 1

Isotopenpraxis **20** (1984) 8, S. 293 bis 296

Reverse Triiodothyronine in Protein Energy Malnutrition

A. A. Hafiez, E. Abdel-Salam, E. Z. Abbas, F. A. Halawa, N. El-Hefnawy

(Biochemistry and Pediatrix Departments, Faculty of Medicine, Cairo University, Cairo, Egypt)

Serum levels of thyroxine (T_4), triiodothyroxine (T_3), reverse triiodothyronine (rT_3) and thyrotropin (TSH) were determined in cases of Kwashiorkor and Marasmus. Decreased levels of T_4 and T_3, and increased levels of rT_3 with no change in TSH were obtained. So, one can conclude that in infants suffering from protein energy malnutrition (PEM) there is a state of thyroid dysfunction and there is a shift in the peripheral T_4 metabolism being converted to the inert rT_3 rather than to the physiologically active T_3.

In Kwashiorkor und Marasmus wurden die Niveaus von Thyroxin (T_4), Trijodthyroxin (T_3), Umkehr-Trijodthyronin (rT_3) und Thyrotropin (TSH) ermittelt. Es wurden verringerte T_4- und T_3-Spiegel, aber erhöhter rT_3-Spiegel festgestellt, während der TSH-Spiegel unverändert blieb. Es kann daher festgestellt werden, daß bei Kindern, die an Proteinmangel (PEM) leiden, eine Funktionsstörung der Schilddrüse vorliegt und eine Verschiebung im peripheren T_4-Stoffwechsel eintritt, weil dies eher in das inerte rT_3 umgewandelt wird als in das physiologisch aktive T_3.

Keywords

children; nutritional deficiency; pediatrics; thyroglobin; thyroid hormones; TRH; triiodothyronine

Introduction

In 1970, it was the first time to demonstrate conclusively that 3,3,5-triiodothyronine is produced by deiodination of the outer phenolic ring of thyroxine [1]. In man, it is now clear that about one third of T_4 metabolism proceeds through the pathways that result in the generation of T_3, a product which is metabolically more active than the parent T_4 [2–4]. They demonstrated that most or nearly all of the circulating T_3 is derived from peripheral T_4 metabolism and not by thyroid secretion. An alternative pathway for peripheral T_4 metabolism is its conversion to 3,3,5-triiodothyronine (reverse T_3 or rT_3), where T_4 is monodeiodinated in its inner ring [5]. Reverse T_3 is considered to be hormonally inactive [6, 7].

It was demonstrated that thyroxine deiodination is depressed during starvation in several species [8, 9]. In man serum concentrations of T_3 were reduced in caloric deprivation and were increased when the subjects were refed [10, 11]. In contrast to T_3, elevated levels of rT_3 were detected in starvation and severe malnutrition which fall upon re-feeding [11, 12]. Also, several studies suggested the presences of primary hypothyroidism in PEM. Thyroid radioiodine uptake is reduced and serum thyrotropin (TSH) and serum TSH response to synthetic thyrotropin releasing hormone (TRH) are elevated in many cases [13–15]. Additionally, thyroid hypoplasia [16] and reduced oxygen consumption [13, 17] are revealed in malnutrition and kwashiorkor in children.

This work was conducted to study the thyroid functions in protein energy malnutrition (Kwashiorkor and Marasmus) by measuring the serum levels of T_4, T_3, rT_3 and TSH.

Materials and Methods

Fifty six children have been studied and were classified into three groups.

The first group included 18 cases of malnutrition (11 male and 7 females), whose age ranged between 6 to 18 months presented with well defined and fully established clinical features of Kwashiorkor.

The second group included 22 cases of malnutrition (10 males and 12 females), their age ranged from 6 to 18 months, and presented with classical picture of Marasmus.

The third group is the control group and comprised 16 well nourished age matched normal infants (8 males and 8 females). Each case was subjected to complete clinical examination specially for symptoms and signs of malnutrition or undernutrition i. e. height & body weight, skin & hair changes, loos of subcutanous fat, liver size and oedema.

Blood samples were taken from all infants, allowed to clot and separated serum kept frozen at −20 °C. The serum levels of T_4, T_3, rT_3 and TSH were estimated in each sample by radio immunoassay kits produced by Serono-Biodata Diagnostica ITALY.

The results are presented in Tab. 1–3 and Fig. 1 and 2.

Analysis of Results and Discussion

A significant decrease is observed in circulating T_4 in Kwashiorkor (mean ± S.D., 102·88 ng/ml ± 35·87, $P < 0{\cdot}05$) and Marasmus (mean ± S.D., 95·55 ng/ml ± 46·32, $P < 0{\cdot}02$) when compared to normal controls (mean ±S.D., 133·63 ng/ml ± 11·22). Similarly, T_3 levels were lower in Kwashiorkor (mean ± S.D., 1·14 ng/ml ± 0·54, $P < 0{\cdot}01$) and Marasmus (mean ± S.D., 1·09 ng/ml ±0·45, $P < 0{\cdot}005$) relative to control group (mean ± S.D., 1·69 ng/ml ± 0·6). The drop in T_4 and T_3 levels were more or less proportional as no significant changes were detected in T_3/T_4 ratio. This drop in both hormones may be explained by the presence of hypothroidism, which was detected in protein energ malnutrition (PEM) by many investigators [13–17]. The possibility that this state of thyroid dysfunction is secondary to pituitary and/or hypothalamic affection may be suggested, as we did not find significant increase in TSH levels. This suggestion is in agreement with those detected low serum levels and low pituitary content of TSH in PEM [18]. A delayed pituitary response to the injection of thyrotropin releasing hormone (TRH) was also observed in infants suffering from malnutrition

Tab. 1. Levels of T_4, T_3, rT_3, T_3/T_4, rT_3/T_3, rT_3/T_4 and TSH in normal controls

	T_4 ng/ml	T_3 ng/ml	rT_3 ng/ml	T_3/T_4	rT_3/T_3	rT_3/T_4	TSH uIU/ml
1	90	1·25	0·2	0·014	0·160	0·0022	2·6
2	170	1·8	0·29	0·011	0·161	0·0017	1·8
3	135	1·4	0·38	0·010	0·271	0·0028	1·5
4	160	1·8	0·29	0·011	0·161	0·0018	1·8
5	195	1·3	0·15	0·007	0·115	0·0008	2·5
6	130	1·2	0·22	0·009	0·183	0·0017	2·6
7	88	1·25	0·08	0·014	0·064	0·0009	2·0
8	130	1·55	0·15	0·001	0·097	0·0012	2·5
9	130	1·55	0·24	0·001	0·155	0·0019	2·6
10	210	2·8	0·1	0·013	0·036	0·0005	1·7
11	95	2·3	0·24	0·024	0·104	0·0025	1·1
12	100	0·9	0·03	0·009	0·031	0·0003	2·7
13	75	0·9	0·1	0·012	0·111	0·0013	3·9
14	200	2·7	0·11	0·014	0·041	0·0006	2·7
15	70	2·3	0·03	0·033	0·013	0·0004	4·9
16	160	2·1	0·17	0·013	0·081	0·0011	3·9
Mean	133·63	1·69	0·17	0·012	0·112	0·0014	2·55
S.D.	35·45	0·60	0·10	0·008	0·068	0·0008	0·99
S.E.	11·3383	0·1492	0·0248	0·0019	0·0171	0·0002	0.2465

Tab. 2. Levels of T_4, T_3, rT_3, T_3/T_4, rT_3/T_3, rT_3/T_4 and TSH in KWO

	T_4 ng/ml	T_3 ng/ml	rT_3 ng/ml	T_3/T_4	rT_3/T_3	rT_3/T_4	TSH uIU/ml
1	148	1·2	0·23	0·008	0·192	0·0016	1·8
2	66	0·95	0·19	0·014	0·2	0·0029	3·0
3	130	0·98	0·64	0·008	0·643	0·0049	1·8
4	96	0·9	0·48	0·009	0·533	0·005	1·4
5	143	1·85	0·33	0·013	0·178	0·0023	4·0
6	130	1·6	0·5	0·012	0·313	0·0039	1·8
7	125	2·0	0·27	0·016	0·135	0·0022	1·9
8	60	0·69	0·32	0·012	0·464	0·0053	1·5
9	130	1·0	0·38	0·008	0·38	0·0029	1·9
10	5	0·82	0·38	0·013	0·463	0·0059	2·0
11	96	1·2	0·21	0·013	0·175	0·0022	2·2
12	75	0·6	0·99	0·008	1·583	0·0127	1·8
13	100	0·8	0·25	0·008	0·313	0·0025	2·2
14	170	2·0	0·14	0·012	0·07	0·0008	2·8
15	119	2·1	0·8	0·018	0·381	0·0067	2·0
16	55	0·5	0·1	0·009	0.20	0·0018	6·0
17	49	0·75	0·1	0·015	0·133	0·002	2·7
18	95	0·57	1·04	0·006	1·825	0·011	1·4
Mean	102·88	1·14	0·41	0·011	0·455	0·0043	2·4
S.D.	35·87	0·5r	0·28	0·003	0·482	0·0032	1·11
S.E.	8·45	0·1260	0·0664	0·001	0·1136	0·0008	0·2614
t	2·17	2·8164	3·3860	0·1331	2·9857	3·5168	0·4175
P	<0·05	<0·01	<0·005	<0·9	<0·005	<0·001	<0·7

which points to a hypothalamic-pituitary axis dysfunctions [19]. Moreover, the role of thyroxine binding proteins in plasma should be considered as many reports have demonstrated a decrease in the plasma levels of thyroxine binding globulin (TBG), thyroxine binding prealbumin (TBPA) and albumin [20—25].

In contrast to both T_4 and T_3, we found an increase in rT_3 levels in Kwashiorkor (mean ± S.D., 0·41 ng/ml ± 0·28, $P < 0{\cdot}005$) and marasmus (mean ± S.D., 0·33 ng/ml ± 0·25, $P < 0{\cdot}01$) relative to normals (mean ± S.D., 0·17 ng/ml ± 0·1). The decrease in T_3 levels and increase in rT_3 levels in these patients indicate that there is a shift in the peripheral T_4 metabolism from generation of the physiologically active T_3 toward the inactive rT_3. This change is further confirmed by the increase in rT_3/T_3 and rT_3/T_4 ratios detected in our cases. These findings are supported by previous reports of may investigators [8 bis 12].

In most of our case, the liver was enlarged and liver affection was reported in PEM [24]. Since liver is one of the major sites for conversion of T_4 to T_3 [26] therefore liver impairment may be a factor responsible for the changes

Tab. 3. Levels of T_4, T_3, rT_3, T_3/T_4, rT_3/T_3, rT_3/T_4 and TSH in MARASMUS cases

	T_4 ng/ml	T_3 ng/ml	rT_3 ng/ml	T_3/T_4	rT_3/T_3	rT_3/T_4	TSH uIU/ml
1	98	0·87	0·82	0'009	0·943	0·0084	2·5
2	38	0·28	0·9	0·007	3·214	0·0237	7·5
3	115	1·55	0·4	0·014	0·258	0·0035	2·6
4	135	1·7	0·38	0·013	0·224	0·0028	1·5
5	90	0·95	0·23	0·011	0·337	0·0036	1·5
6	80	1·3	0·22	0·016	0·169	0·0028	2·0
7	150	1·5	0·4	0·01	0·267	0·0027	2·0
8	18	0·47	0·18	0·026	0·383	0·01	7·5
9	90	0·95	0·15	0·01	0·158	0·0017	2·6
10	95	0·74	0·36	0·008	0·486	0·0038	1·8
11	35	1·25	0·4	0·035	0·32	0·0114	2·7
12	140	1·9	0·32	0·0014	0·168	0·0023	2·0
13	90	1·05	0·23	0·012	0·219	0·0026	1·6
14	95	2·1	0·85	0·022	0·405	0·0090	1·5
15	70	0·85	0·03	0·012	0·025	0·0043	1·8
16	55	0·9	0·01	0·16	0·017	0·0036	1·3
17	80	0·8	0·55	0·01	0·688	0·0069	1·1
18	150	0·8	0·16	0·005	0·2	0·0011	1·9
19	220	0·9	0·11	0·004	0·122	0·0005	3·7
20	130	1·05	0·14	0·008	0·133	0·0011	2·9
21	48	0·7	0·14	0·014	0·2	0·0029	5·5
22	80	1·25	0·15	0·015	0·12	0·019	4·9
Mean	95·55	1·09	0·33	0·013	0·412	0·0060	2·88
S.D.	46·23	0·45	0·25	0·007	0·661	0·0051	1·84
S.E.	9·86	0·0957	0·054	0·0015	0·1408	0·0011	0·6144
t	2·54	3·3850	2·6926	0·4131	2·1151	3·2199	0·4985
P	<0·02	<0·005	<0·01	<0·7	<0·05	<0·005	<0·7

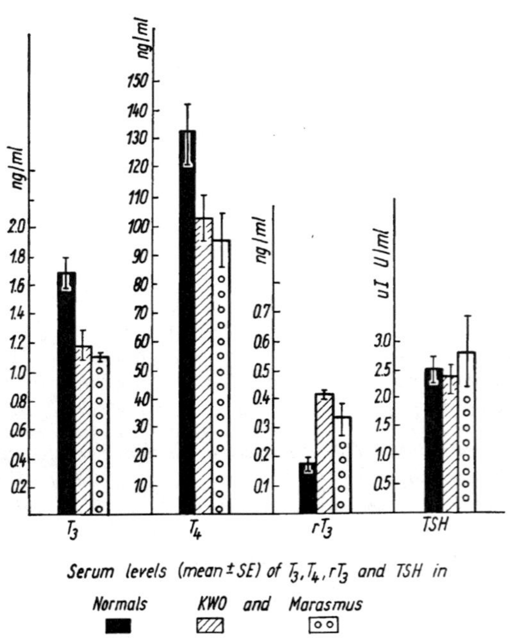

Fig. 1

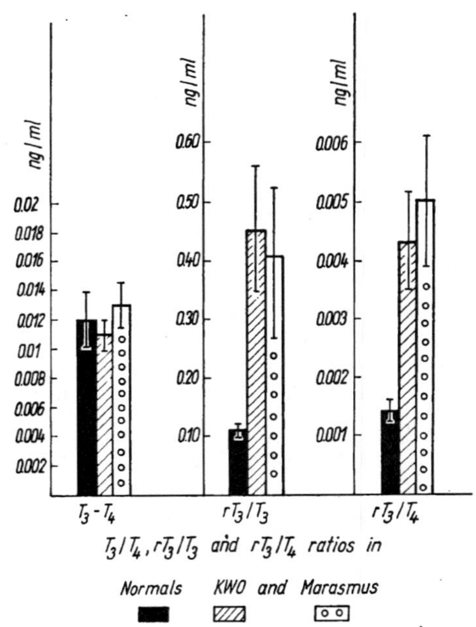

Fig. 2

detected in PEM. This suggestion is in agreement with the finding that there is decrease in plasma T_3 levels in liver diseases [27, 28].

In addition, the possibility exists that carbohydrate deficiency in marasmus plays an important role as it was stated that dietary carbohydrate is an important regulatory factor in T_3 production in man [29]. Also, many reports have detected a lowered levels of T_3 in diabetics where glucose utilization is decreased [30–32]. Recently it is proved that glucose is an important cofactor for the liver microsomal enzyme responsible for conversion of T_4 to T_3 [33]. They postulated that the 5 monodeiodinase for thyroxine that mediates T_3 neogenesis may be glutathione transhydrogenase which requires NADPH and lateral could be supplied by glucose utilization through the hexose monophosphate pathway (HMP).

It is of interest to note that deficiences of certain vitamins may be associated with a decrease in hepatic deiodination of thyroxine [34, 35]. Since vitamin deficiences occur frequently with malnutritions, so this may be another factor responsible for the decreased T_3 levels encountered.

However, it is stated that the diversion of T_4 conversion from the highly potent T_3 towards the generation of poorly calorigenic rT_3 may be a defence reaction of the body attempting to protect sick tissues from excessive metabolic stimulation [12].

It can be concluded, from this study, that in infants suffering from PEM there is a state of thyroid dysfunction which is pressented by a low level of serum T_4 and T_3, elevated level of rT_3 and no change in TSH.

Received March 16, 1983
Accepted in revised form November 2, 1983

References

[1] *L. E. Braverman, S. H. Ingbar, K. Sterling*, J. Clin. Invest. **49** (1970) 855
[2] *C. S. Pittman, J. B. Chambers*, Jr., *V. H. Read*, J. Clin. Invest. **50** (1971) 1187
[3] *M. I. Surks, A. R. Schadlow, J. M. Stock, J. H. Oppenheimer*, J. Clin. Invest. **52** (1973) 805
[4] *M. K. Inada, S. Kasagi, Kurata, Y. Kazama*, J. Clin. Invest. **55** (1975) 1337
[5] *I. J. Chopra*, J. Clin. Invest. **58** (1976) 32
[6] *W. L. Money, S. Kumar, R. W. Ramson, R. L. Kroc*, Ann. N. Y. Acad. Sci **86** (1960) 512
[7] *S. B. Barker, M. Shimada, M. Makuichi*, Endocrinology **76** (1965) 115
[8] *S. Reichlin*, Endocrinology **60** (1957) 470
[9] *D. Ingbar, V. Galton*, Endocrinology **96** (1975) 1525
[10] *G. I. Portnay, J. T. O'Brian, A. G. Vagenakis, F. Azizi, R. A. Arky, S. H. Ingbar, L. E. Braverman*, J. Clin. Endocr. Metab. **39** (1974) 191
[11] *A. G. Vagenakis, A. Burger, G. I. Portany, M. Rudolph, J. T. O'Brian, F. Azizi, R. A. Arky, P. Nicod, S. H. Ingbar, L. E. Braverman*, J. Clin. Endocr. **41** (1975) 191
[12] *I. J. Chopra, U. Chopra, S. R. Smith, M. Reza, D. H. Solomon*, J. Clin. Endocr. Metab. **41** (1975) 1043
[13] *F. Varga*, Pediatrics **23** (1959) 1085
[14] *F. Beas, F. Monchegery, I. Horwitz, M. Figueraa*, Pediatrics **38** (1966) 1003
[15] *B. Pimstone, D. Becker, S. Hendricks*, J. Clinl. Endocr. Metab. **36** (1973) 779
[16] *C. Tajada*, Ped. Proc. **17** (1958) 66
[17] *J. C. Ablett, R. A. McCance*, Lancet **2** (1971) 517
[18] *F. Varga, B. Mess*, Pediatr. Acad. Sci. Hung. **9** (1968) 197
[19] *F. A. El-Grazzar*, M. D. Thesis, Faculty of Medicine, Cairo University (1980)
[20] *A. A. Ismail, Abdel-Hay*, Gaz. Egypt. Ped. Assoc. **3** (1960) 475
[21] *A. Sawad, E. M. Abdel-Wahab*, Ain Shams Med. J. **13** (1962) 42
[22] *S. Awad, E. M. Abdel-Wahab*, Ped. Assoc. XIV (1966) 160
[23] *S. Cohen, J. D. L. Hansen*, Clin. Sci. **23** (1962) 351
[24] *A. H. Shehata, A. Abdel-Hay, G. Kamel, B. Kammah, I. Fayed, M. Safouf, M. Talaat*, J. Egypt. Endocr. Metab. **11** (1965) 11
[25] *Y. Ingenbleek, M. de Visscher, ph. de Nayer*, Lancet **2** (1972) 106
[26] *A. Balsam, S. H. Ingbar, P. Sexton*, J. Clin. Invest. **62** (1978) 415
[27] *I. J. Chopra, D. H. Solomon, U. Chopra, R. T. Yound, G. N. Chua Teco*, J. clin. Invest. **54** (1974) 583
[28] *S. Nomura, C. S. Pittman, J. B. Chambers*, Jr. *M. W. Buck, T. Shimza*, J. Clin. Invest. **56** (1975) 643
[29] *S. W. Spandling, I.J. Chopra, R. S. Sherwin, S. S. Lyl*, J. Clin. Endocr. Metab. **42** (1976) 197
[30] *R. Nacije, N. Clumeck, G. Somers, L. Vanhealst, J. Golstein*, Lancet **1** (1976) 1070
[31] *E. Pirosk, J. Foldes*, Lancet **1** (1967) 1071
[32] *A. A. Hafiez, A. A. Ismail, M. R. Awadeen, E. Z. Abbas, M. S. Farag*, the third scientific conference, Egyptian Society of Basis Medical Sciences, 1981
[33] *A. Balsam, H. Sidney*, J. Clin. Invest. **63** (1972) 1145
[34] *V. A. Galton, S. H. Ingbar*, Endocrinology **77** (1965) 169
[35] *J. M. Hsu*, J. Nutr. **97** (1965) 505

Isotopenpraxis **20** (1984) 8, S. 296 bis 299

Effect of Estrogens on Urinary ^{15}N-Balance in Girls*)

M. Zachmann, B. Kempken, A. Prader

(Department of Pediatrics, University of Zürich, Kinderspital)[1]

While the anabolic and growth-promoting effects of testosterone (T) are known to be important for pubertal growth in boys, the role of estrogens (E) in the female spurt is less certain. Adrenal androgens have been considered to be more important than ovanian E. To study the anabolic effects of E, a pilot study in 9 girls aged 11 to 15 years was carried out. Before and 6 days after the start of E-treatment, urinary ^{15}N-balance studies were performed, using $^{15}NH_4Cl$.

Während die anabolische und wachstumsfördernde Wirkung des Testosterons (T) als wichtig für das Pubertätswachstum bei Jungen bekannt ist, ist man sich hinsichtlich der Rolle der Östrogene (Ö) beim weiblichen Wachstumsschub weniger sicher. Es wurde angenommen, daß Nebennierenandrogene wichtiger sind als das ovarielle Ö. An 9 Mädchen im Alter von 11 bis 15 Jahren wurde eine Pilotuntersuchung durchgeführt, um den anabolen Einfluß der Östrogene zu ermitteln. Vor Beginn der Östrogen-Behandlung und 6 Tage danach wurden unter Verwendung von $^{15}NH_4Cl$ ^{15}N-Bilanzen untersucht.

Keywords

estrogens; nitrogen 15; pediatrics; pubertal growth; STH

In previous studies, we have several years ago been interested in the response of urinary nitrogen excretion to human growth hormone (hGH)-treatment in children with short stature as a biological test for growth hormone deficiency [1].

The practical difficulties with these tests, which require a constant diet and hospitalization for 2 weeks, prompted us to try ^{15}N-labelled compounds for short term nitrogen balance tests.

---

*) Presented at the "^{15}N-Symposium", 29th May to 1st June 1983, Rostock.

[1]) Address: 8032-Zürich, Switzerland, Steinwiesenstraße 75

For this purpose, the subjects were given 0·1 g of 50% ^{15}N-labelled ammonium chloride per kg of body weight, once before and once after hormone stimulation. The ^{15}N-labelled NH_4Cl was obtained from VEB Berlin-Chemie, Berlin-Adlershof (GDR). On the test days, the NH_4Cl was diluted in water (40 ml per g of NH_4Cl) and fruit juice (20 ml per g of NH_4Cl) and divided into 3 equal doses to be taken by mouth at 8 a.m., 12 noon, and 4 p.m. Starting at 8 a.m., a 24-h urine specimen was collected. The NH_4Cl dose corresponded to 13·4 to 15·5.mg of ^{15}N per kg (mean 14·2 ± 0·8 mg/kg) or to 331 to 500 mg per square meter of body surface area (mean 459 ± 49 mg/m²).

All subjects were ambulatory during the test period and were on their regular diet.

The NH_4Cl was well tolerated, although its taste was unpleasant, especially to younger children. With the exception of a mild metabolic acidosis towards the end of the test days (base excess −4 to −6 mequiv./l at 6 p.m.), no side-effects were observed.

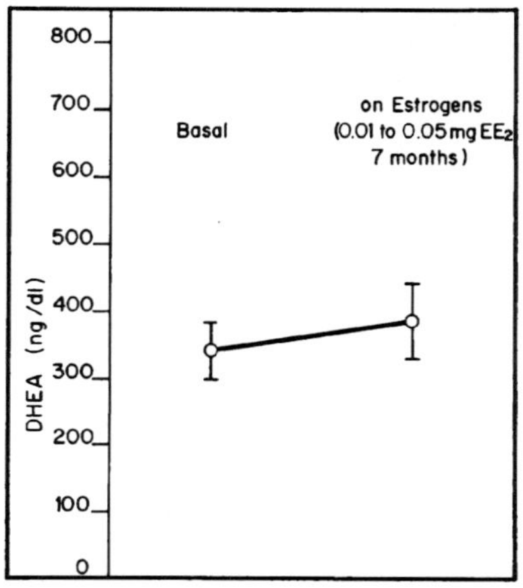

Fig. 1. Plasma dehydroepiandrosterone in 8 girls with Turner-syndrome before and on estrogen replacement therapy
ethinyl estradiol 0·01 to 0·05 mg daily

The samples were prepared according to a modified *Kjeldahl* method in the following way: 10 ml of urine from the well-mixed 24-h sample were reduced to a volume of about 1 ml on a rotary evaporator. Depending on the 24-h volume and the age of the subject, a larger or smaller quantity of urine may be required; 8 ml of selenic sulfuric acid (0·7 g SeO_2 and 100 ml of 47·5 to 48·5% H_2SO_4), 10 drops of 10% (v/v) copper sulfate solution and 25 ml of 33% NaOH (for distillation) were used instead of the quantities of 4 ml, 5 drops and 20 ml respectively as in the regular *Kjeldahl* procedure; instead of titrating the distillate, an acid excess (3 to 5 drops of 2,5 M H_2SO_4) was then added.

The distillate was collected in a recipient containing an aqueous solution of the mixed indicator No. 5 (Merck No. 6130) instead of *Tashiro*'s reagent. After titration with 0·005 M H_2SO_4 (dark-purple color), the amount of acid used was recorded and an excess (3 ml of 0·005 M H_2SO_4) was added. Thereafter, the titrated distillate was dried on a rotary evaporator and the residue was diluted in bidistilled water (1/5 of the 0·005 M H_2SO_4 volume used for titration). 0·2 ml of this solution were then pipetted into one chamber of a special recipient according to *Rittenberg*, with 0·2 ml of alkaline hypobromide solution (200 g NaOH in 300 ml bidistilled water +60 ml of bromide p.a.) pipetted into the other chamber. The recipient was then connected with a special high-vacuum generating apparatus. Air and water vapor were removed by repeated freezing and thawing and the oxidation of the sample with NaOBr to molecular nitrogen was subsequently carried out unter vacuum at 10^{-5} Torr. The discharge tubes were filled with the nitrogen gas liberated during the reaction and the collected gas was then trapped in the tubes by fusing the glass end. From each sample, 2 to 3 tubes were prepared. With the content of these tubes, the spectroscopic isotope analysis was carried out on an Isocommerz model NO15 ^{15}N-analyzer (VEB Statron, Fürstenwalde, GDR 2 to 6) scanned between 2977 to 2988 Å [2—6]. The mean of 10 scans was taken to calculate the ^{15}N-percentage by com-

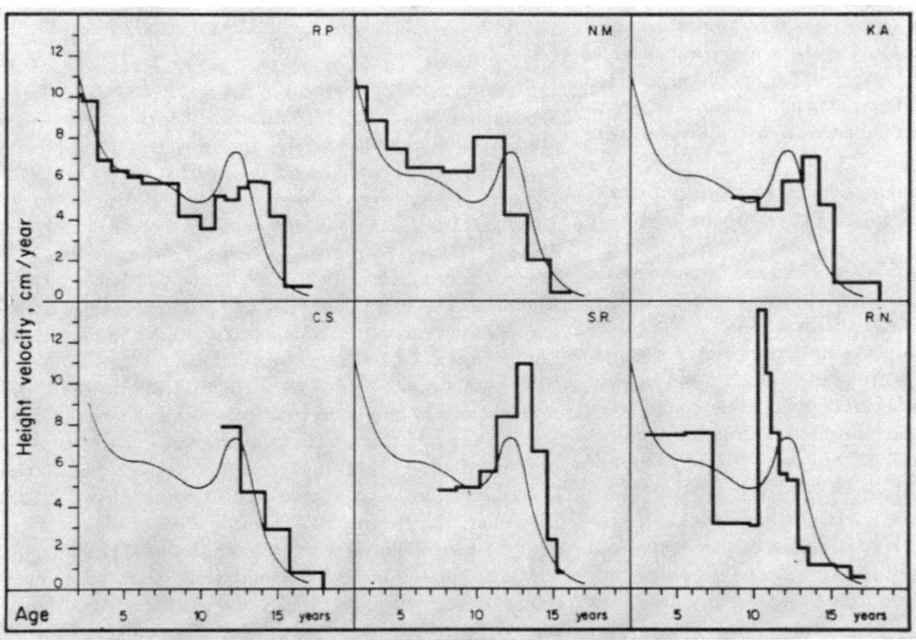

Fig. 2. Pubertal growth spurt in 6 patients with the syndrome of testicular feminization
5 patients had no tratment. 1 patient (lower right panel) was gonadectomized in early childhood, and was treated with conjugated estrogens at a pubertal age.

parison with a calibration curve. The results were corrected for the relative frequency of the atmospheric isotope concentration (0·366%). With this test design, we observed that patients with growth hormone deficiency responded with a more positive urinary ^{15}N-balance to hGH (2 daily im injections of 2·6 IU/m^2 each) than normal children [7], and that the response was modified by testosterone [8].

In another series of studies, we saw that there was a positive ^{15}N-balance change also in response to testosterone (15 mg/m^2 of testosterone propionate im daily for 6 days) in prebupertal boys, but not in patients with the syndrome of testicular feminization [9, 10].

While the pubertal growth spurt in boys is clearly a consequence of the protein-anabolic action of testosterone in the presence of growth hormone and thyroxine [11], the situation is by far less clear in girls, and there is still considerable controversy concerning the cause or causes of their pubertal spurt.

An opinion to be read in textbooks is that estrogens do not accelerate growth in girls, but that adrenal androgens such as dehydroepiandrosterone and androstenedione are responsible for the phenomenon.

Recent studies concerning adrenarche, however, show that the timing does not support this concept. DHEA in plasma starts to increcse quite early, at about 7 or 8 years of age, while the peak of pubertal growth in girls is reached after 11 years only. Another theory says that the rise of estrogens during puberty has itself an increasing effect on adrenal androgen secretion. While there is some in vitro evidence in favour of this hypothesis, in vivo studies do also not support it.

We had recently the opportunity to study plasma DHEA in 8 girls with Turner syndrome, before and after estrogen replacement therapy, and there was no change (Fig. 1). It is therefore unlikely that adrenal androgens play a major role in the pubertal growth spurt of girls, and a direct effect of estrogens seems to be more probable. Another observation supports this view: we could follow the pubertal growth of 6 girls, with the syndrome of testicular feminization, where an action of androgens is excluded. The first 5 patients had no treatment, and had nevertheless a very marked pubertal growth acceleration (Fig. 2). The sixth patient, had been gonadectomized in early childhood, and was treated with conjugated estrogens at a pubertal age, which caused a marked acceleration of the growth rate.

These clinical experiences support the assumption of the growth promoting, anabolic effect of estrogens alone in girls.

For these reasons, we planned to examine, whether also with ^{15}N-studies, a positive balance change could be induced by estrogens. This was indeed the case. Since this project is quite recent, we have only a small number of results. So far, we have carried out ^{15}N-balance studies before and after estrogen administration in 6 girls. The tests were done in the same way with enriched ammonium chloride, as previously described. They were performed before, and 5 days after initiation of estrogen treatment.

Three patients were normal girls with familial tall stature, in whom treatment with high doses of estrogens was given to reduce their final adult height. They received an im-injection of 40 mg of estradiolvalerate every 2 weeks or ethinylestradiol 0·3 mg daily by mouth, and the second test was performed on the 6th day after initiation of therapy. Three others were girls with gonadal dysgenesis and various sex chromosome anomalies. They received estrogen replacement doses (ethinyl estradiol 0·01 to 0·05 mg daily by mouth).

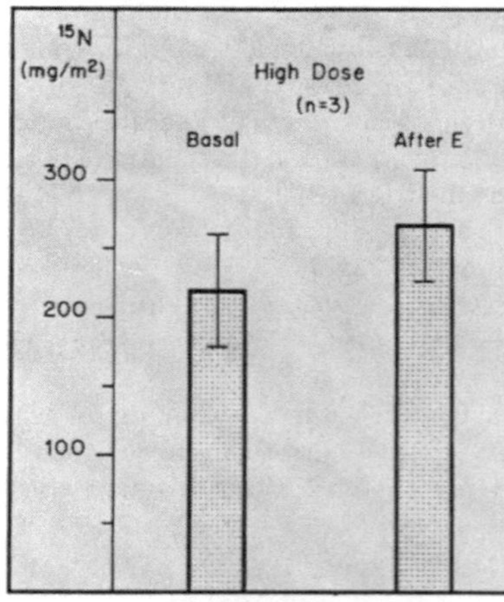

Fig. 3. Urinary ^{15}N-balance (expressed in mg ^{15}N/m^2 of body surface area) in 3 tall girls before and on high dose estrogen treatment

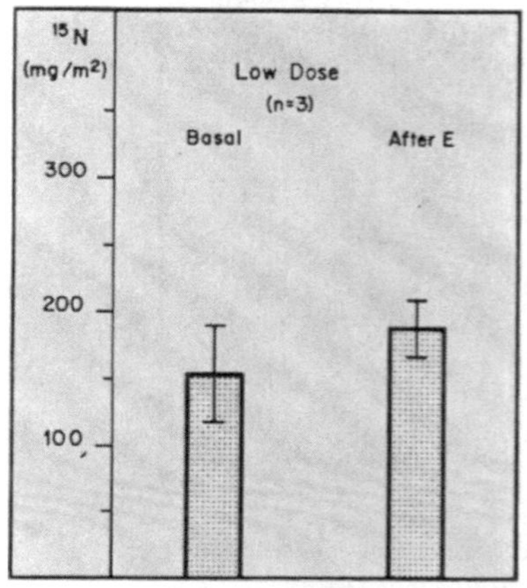

Fig. 4. Urinary ^{15}N-balance (mg ^{15}N/m^2) in 3 girls with gonadal dysgenesis before and on low dose estrogen replacement

The basal ^{15}N-balance was with +219 mg/m^2 slightly higher in the tall girls than in those with gonadal dysgenesis, whose balance was +153 mg/m^2. Both groups showed a slight positive balance change on estrogen treatment. In the tall girls, the balanced changed from +219 to 267 mg per m^2, which corresponds to 22% of the basal balance (Fig. 3). In the girls with gonadal dysgenesis, it changed from +153 to 188 mg/m^2, which corresponds to 23% of the pretreatment balance (Fig. 4). Thus, the basal balance was higher in the tall girls with the higher endogenous estrogen production, but the balance change after estrogens was about equal in both groups inspite of the fact that the estrogens dose in the tall girls was about 6 to 10 times higher than in the girls with gonadal dysgenesis. The number of cases in this study is small, and the results have to be regarded as preliminary ones, since the population was heterogeneous with respect to diagnosis and treatment. However, the results fit our clinical expectation that estrogens do indeed have a protein anabolic effect in girls, but that this effect is less marked than that of growth hormone or testosterone. Again the previous experience with growth hormone and testosterone is confirmed, which indicates that there is a response to a smaller quantity of hormone, when the endogeneous hormone production is low.

Acknowledgement

This work was supported by the Swiss National Science Foundation (Grants No. 3.959.080, and 3.969.180).

Received October 26, 1983

References

[1] *A. Prader, M. Zachmann, J. R. Poley, R. Illig*, Acta Endocrinol. (Kbh.) 57 (1968) 115

[2] *K. Wetzel, H. Faust, W. Hartig*, in Proceedings of the Second International Conference on Stable Isotopes (*E. R. Klein*, and *P. D. Klein*, eds.), p. 400, The National Technical Information Service, US Dept. of Commerce, Springfield, Va. (1976)

[3] *G. Meyer, G. Muller*, Isotopenpraxis 1 (1965) 53

[4] *I. P. Leicknam, V. Middelboe, G. Proksch*, Anal. Chim. Acta 40 (1968) 487

[5] *K. Kumazawa*, Radiositopes 21 (1972) 623

[6] *H. Gerstenberger, G. Meyer, K. Mauersberger, G. Müller, D. Pabst, S. Riedel, W. Rose, G. Schmidt*, GIT-Fachz. Labor 18 (1974), 452

[7] *M. Zachmann, J. A. Völlmin, A. Prader* Acta Paediatr. Scand. 62 (1973) 97

[8] *M. Zachmann, M. Zagalak, R. P. Gitzelmann, A. Prader*, in: *E. R. Klein* and *P. D. Klein* (eds.) Stable Isotopes, New York: Academic Press 1979, p. 619

[9] *M. Zachmann, M. Zagalak, J. A. Völlmin, A. Prader* in Proceedings of the Second International Conference on Stable Isotopes (*E. R. Klein* and *P. D. Klein*, eds.), p. 374, The National Technical Information Service, US Dept. of Commerce, Springfield, Va. (1976)

[10] *M. Zachmann, M. Zagalak, J. A. Völlmin, R. P. Gitzelmann, A. Prader*, Clin. Chim. Acta 77 (1977) 147

[11] *A. Aynsley-Green, M. Zachmann, A. Prader*, J. Pediat. 89 (1976) 992

Isotopenpraxis 20 (1984) 8, S. 299 bis 300

Zur Retention von ^{60}Co, ^{58}Co/^{54}Mn und ^{110m}Ag nach inhalativer Aufnahme im Menschen

K.-F. Poulheim

(Staatliches Amt für Atomsicherheit und Strahlenschutz)[1])

Es wird ein Beitrag zur Retention der crud-Radionuklide ^{60}Co, ^{58}Co/^{54}Mn und ^{110m}Ag im Menschen nach inhalativer Aufnahme gegeben. Die Untersuchungen wurden an 10 Personen, die über die Inkorporationsüberwachung erfaßt wurden durchgeführt. Die ermittelten effektiven Halbwertszeiten der genannten Radionuklide werden mit Literaturwerten verglichen.

A contribution on the retention of the crud radionuclides ^{60}Co, ^{58}Co/^{54}Mn and ^{110m}Ag in human beings after incorporation by inhalation is presented. The investigations were carried out on 10 persons subject to incorporation control. The obtained effective half-life values of the said radionuclides are compared to values found in literature.

Keywords

biological half-life; cobalt 58; cobalt 60; inhalation; intake; manganese 54; medical surveillance; retention function; silver 110m

Die ICRP beschäftigt sich seit längerem mit einer präzisen Erfassung aller erhältlichen Daten über die Retention von Radionukliden. In diesem Zusammenhang ist es wichtig, alle Untersuchungen an Personen auszuwerten, die in der Folge beruflicher Strahlenexposition Radionuklide inkorporiert haben. Dank der heute hochentwickelten Ganzkörperzähler ist es bereits möglich, sehr kleine inkorporierte Aktivitäten, deren Dosisberechnung für die betreffende Person belanglos geringe Äquivalentdosen ergibt, hinreichend lange meßtechnisch zu verfolgen.

Als ein Ergebnis der Inkorporationsüberwachung mit dem Ganzkörperzähler des Staatlichen Amtes für Atomsicherheit und Strahlenschutz sind die hier dargestellten Retentionsverläufe von ^{60}Co, ^{58}Co/^{54}Mn und ^{110m}Ag in insgesamt 10 beruflich strahlenexponierten Personen zu werten. Die Abschätzungen der inneren Strahlenbelastung ergaben in allen Fällen Resultate, die etwa 1% der Äquivalentdosis durch äußere Strahlenbelastung betrugen. Die Fallbeschreibungen wurden bereits früher ausgeführt [1].

Im Ganzkörperzähler [2, 3] wurden neben der Gesamtaktivitätsbestimmung für die einzelnen Radionuklide mit Hilfe der NaJ(Tl)-γ-Spektrometrie longitudinale Scan-Messungen mit einem Schlitzkollimator (Schlitzbreite 1 – 2 cm) entlang der Körperachse der Personen sowie γ-spektrometrische Messungen mit einem Ge(Li)-Detektor am Thorax vorgenommen. Bei allen Personen, die in die Retentionsuntersuchungen einbezogen wurden, war die gesamte oder wenigstens mehr als 80% der Aktivität im Thoraxbereich zu lokalisieren.

Aus den bestimmten effektiven Halbwertszeiten T_{eff} für ^{60}Co nach Inhalation (wahrscheinlich in Form von Cobaltoxid) kann folgende Retentionsgleichung aufgestellt werden:

$$R(t) = 0{,}2 \cdot e^{\frac{-0{,}693}{42} t} + 0{,}8 \cdot e^{\frac{-0{,}693}{98} t}$$

mit t in Tagen $T_{\text{eff}} = (42 \pm 14)$ d gilt vom 7. bis 40. Tag nach dem Inkorporationszeitpunkt und $T_{\text{eff}} = (98 \pm 26)$ d für die Zeit vom 40. bis etwa 120. Tag (Tab. 1). Eine schnelle Clearance-Komponente von einigen Tagen konnte nur in einem Fall bestimmt werden, da zumeist die ersten Messungen im Ganzkörperzähler erst mehrere Tage nach dem Inkorporationszeitpunkt erfolgten. Die ermittelten T_{eff}-Werte stimmen mit Werten von *Cofield* [4] (90 d), *Sedlet* u. a. [5] (42 d, 110 d), *Jordan* [6] (115 d), *Bhat* u. a. [7] (43 d) und *Hegde* u a. [8] (56 d) überein. Außer diesen mittelschnellen Komponenten wiesen mehrere Autoren auch eine sehr langsame Clearance nach. Im relativ kurzen Untersuchungsintervall dieser Meßreihen von 50 bis 139 Tagen gelang der Nachweis von großen T_{eff}-Werten bis zu mehreren Jahren nicht.

[1]) Anschrift: 1157 Berlin, Waldowallee 117

Tab. 1. Retention von ^{60}Co und $^{58}Co/^{54}Mn$

Intervall zwischen 1. u. letzter Messg. d	^{60}Co effektive Halbwertszeit, i. Klammern der %uale Anteil der Komponente d			$^{58}Co/^{54}Mn$ effektive Halbwertszeit, i. Klammern der %uale Anteil der Komponente d		
123			102 (100)			56 (100)
118			86 (100)			60 (100)
125			91 (100)			53 (100)
139	4 (95)		72 (5)	3 (96)		44 (4)
119		42 (100)			33 (100)	
50		51 (12)	111 (88)		32 (25)	70 (75)
56		27,5 (29)	98 (71)		34 (40)	83 (60)
53		46 (20)	123 (80)		31,5 (24)	49 (76)

Tab. 2. Retention von ^{110m}Ag

Intervall zwischen 1. u. letzter Messg. d	effektive Halbwertszeit, i. Klammern der %uale Anteil der Komponenten d		
71	5,4 (80)	27 (20)	
44	3,2 (75)	22,5 (25)	
57		26 (28)	44 (72)
50		34 (100)	

Aus den Meßreihen kann für die crud-Nuklide $^{58}Co/^{54}Mn$ die Retentionsgleichung

$$R(t) = 0{,}3 \cdot e^{-\frac{0{,}693}{33}t} + 0{,}7 \cdot e^{-\frac{0{,}693}{60}t}$$

mit t in Tagen aufgestellt werden. Dabei sind $T_{eff} = (33 \pm 1)$ d für den 7. bis 40. Tag und $T_{eff} = (60 \pm 23)$ d für den 40. bis 120. Tag nach dem Zeitpunkt der Inkorporation anzugeben.

Ge(Li)-γ-spektrometrische Messungen am Thorax ergaben für ^{58}Co $T_{eff} = (62 \pm 14)$ d und für ^{54}Mn $T_{eff} = (153 \pm 80)$ d. Das Verhältnis von $^{58}Co/^{54}Mn$ betrug zwischen 3:2 bis 6:1, d. h. der Anteil des ^{58}Co ($T_{phys} = 71{,}3$ d) dominierte in allen Fällen.

Die Messungen ergaben für ^{110m}Ag die Retentionsgleichung

$$R(t) = 0{,}80 \cdot e^{-\frac{0{,}693}{4{,}3}t} + 0{,}5 \cdot e^{-\frac{0{,}693}{28}t} + 0{,}15 \cdot e^{-\frac{0{,}693}{44}t}$$

mit t in Tagen. Die effektive Halbwertszeit von 4,3 d gilt in der Zeit vom 1. bis 10. Tag nach dem Inkorporationszeitpunkt, $T_{eff} = 28$ d für die anschließende Periode bis zum 52. Tag und $T_{eff} = 44$ d nach dem 52. Tag (Tab. 2).

Die effektive Halbwertszeit der langsamen Komponente (44 d) stimmt sehr gut mit den Werten von *Newton* und *Holmes* [9] überein (1 d — 84%, 43 d — 16%). Auch die prozentualen Anteile sind gleich.

Eingegangen am 28. 9. 1983

Literatur

[1] *K.-F. Poulheim*, Исследование удержания радионуклидов с помощью измерений на Сиче" RGW-Konferenz Оьеспечение радидц-ионной дезопасности при эксплуатации дэс UdSSR, Vilnius, Mai 1982 (im Druck)

[2] *V. Lößner, K.-F. Poulheim, W. Schlüter, H. Klucke*, Einrichtungen zur physikalischen Inkorporationskontrolle im Staatlichen Amt für Atomicherheit und Strahlenschutz, Report SZS-163, Juli 1974

[3] *K.-F. Poulheim*, Radiol. Radiother. 18 (1977) 263

[4] *R. E. Cofield*, Health Physics 9 (1963) 283

[5] *J. Sedlat*, "Elimination of Radioactive Cobalt Following Acute Accidental Inhalation", IPRA Congress, England, Brighton 1970

[6] *R. D. Jordan, J. S. Burkle, L. T. Brown, J. W. Margus, J. H. Nichols*, "Cobalt^{60} oxide Inhalation", in "Radioactivity in Man", *C. C. Thomas* (ed.), Springfield, Illinois 1965, S. 281

[7] *I. S. Bhat, A. G. Hegde, S. Chandramouli, R. S. Iyer*, Health Physics 25 (1973) 8, S. 135

[8] *A. G. Hegee, D. M. Thakkerk I. S. Bhat*, Health Physics 36 (1979) 6, S. 734

[9] *D. Newton, A. Holmes*, Radiat. Res. 29 (1966) 403

Isotopenpraxis 20 (1984) 8, S. 301 bis 303

Spurenelementkonzentration im Haar bei Patienten mit Schilddrüsenerkrankungen*)

T. Mende[1]), *W. Wiesener*[2]), *W.-G. Franke*[1]), *S. Domschke*,[1]) *W. Görner*[2])

([1]) Medizinische Akademie „Carl Gustav Carus" Dresden, Abteilung für Nuklearmedizin; [2]) Zentralinstitut für Kernforschung Rossendorf)

Mit Hilfe der Neutronenaktivierungsanalyse wurden in mehr als 200 Haarproben Iod, Gold, Zink, Mangan, Kupfer, Brom, Natrium und in einigen Fällen Selen bestimmt (2, 4). Zur Untersuchung kam das Haar von Patienten mit verschiedenen Schilddrüsenerkrankungen: Struma kombiniert mit Euthyreose, erhöhter Iodavidität, Hyperthyreose und Schilddrüsencarcinom. Als Kontrollgruppen wurden Probanden mit Euthyreose sowie ein ausgewähltes Kollektiv, das offensichtlich nicht an einer Schilddrüsenerkrankung litt, herangezogen. Die Ergebnisse zeigen, daß die Bestimmung des Spurenelementstatus im Haar eine Erweiterung der diagnostischen Möglichkeiten in bezug auf Schilddrüsenfunktion und Schilddrüsenerkrankungen bedeutet.

By means of neutron activation analyses the content of iodine, bromine, sodium, gold, zinc, copper, manganese and in some cases of selene was determined in more than 200 hair samples. The study was undertaken on hair of patients with diseases of the thyroid (struma combined with euthyreosis, struma with increased iodine avidity, hyperthyreosis, carcinoma of the thyroid). The control group consisted of test persons with euthyreosis, as well as a random group, that obviously did not suffer from a thyroid disease.

Keywords

activation analysis; carcinomas; copper; epidemiology; hair; iodine; thyroiditis; trace elements; zinc

Material und Methodik

Die Haarproben werden von der Nackenpartie entnommen, nach einer von der IAEA empfohlenen Vorschrift [5] nacheinander mit Aceton, dreimal mit tridestilliertem Wasser und Aceton gewaschen, bei 80 °C 2 h getrocknet und in Tüten aus Polyäthylenfolie oder in Quarzampullen eingewogen. Die Einwaagen liegen zwischen 10 und 40 mg. Die Bestrahlung wird im Rossendorfer Forschungsreaktor durchgeführt. Für die Iod- und Selenbestimmung genügt eine Kurzzeitaktivierung von 15–120 s in einer pneumatischen Bestrahlungsanlage bei einer Neutronenflußdichte $\Phi_{th} = 4 \ldots 6 \cdot 10^{13}$ n cm^{-2} s^{-1}. Für die übrigen Elemente wählt man eine Bestrahlungszeit von 1 h im Reaktorcore bei einer Flußdichte $\Phi_{th} = 6 \ldots 8 \cdot 10^{13}$ n cm^{-2} s^{-1}. Alle Proben werden ohne chemische Aufarbeitung, d. h. zerstörungsfrei nach der Bestrahlung spektrometriert. Die Aktivitätsmessungen werden mit einem 4096-Kanal-Speicher unter Verwendung eines 18 cm^3-Ge(Li)-Detektors (Auflösung 2,8 keV bei 1330 keV) durchgeführt. Zur Spektrometrie der kurzlebigen Radionuklide (^{128}I, 443 keV, $\tau_{1/2} = 25$ min; ^{77m}Se, 162 keV, $\tau_{1/2} = 17{,}5$ s) wird ein automatischer Totzeitkorrektor eingesetzt. Für die Nuklide ^{69m}Zn, ^{64}Cu, ^{56}Mn, ^{24}Na, ^{82}Br, ^{198}Au liegen die Meßzeiten bei 400 s. Die gesuchte Elementkonzentration errechnet sich aus der Intensität der Gammastrahlung durch einen Vergleich mit der Aktivität der mitbestrahlten Standards.

Parallel zur Bestimmung der Spurenelemente erfolgte bei allen Probanden die nuklearmedizinische Diagnostik in Form des Radioiodzweiphasenstudiums und der Schilddrüsenszintigraphie; bei den Carcinompatienten lagen der histologische Befund und der Operationsbefund vor.

Im Vordergrund des medizinischen Interesses im Hinblick auf die Anwendbarkeit stand die Iodbestimmung im Haar, da Iod einen zentralen Platz im Schilddrüsen- und Schilddrüsenhormonstoffwechsel einnimmt.

Ergebnisse und Diskussion

In Übereinstimmung mit eigenen Voruntersuchungen zeigen Patienten mit Struma niedrigere Iodwerte im Haar als die Kontrollgruppe [3]. Der Iodgehalt im Haar von Patienten mit Hyperthyreose lag durchschnittlich höher als bei Strumaträgern mit Euthyreose (Tab. 1). Abb. 1 zeigt die Häufigkeitsverteilungen der Iodwerte im Haar von Patienten mit Euthyreose und Struma, mit erhöhter

Tab. 1. Iodgehalt im Haar (ppm) von Patienten mit verschiedenen Schilddrüsenerkrankungen (Raum Dresden 1977–1982)

Gruppe		*n*	$\bar{x}$	*s*
Kontrollgruppe	♀ und ♂	100	1,20	0,93
	♂	76	1,22	0,69
	♀	24	1,14	1,46
Euthyreose und Struma einschließlich erhöhter Iodavidität der Schilddrüse	♀ und ♂	41	0,31	0,21
	♀	32	0,33	0,16
	♂	7	0,24	0,13
Euthyreose und Struma, erhöhte Iodavidität der Schilddrüse ausgenommen	♀ und ♂	12	0,37	0,32
	♀	8	0,48	0,35
	♂	4	0,16	0,10
Erhöhte Iodavidität der Schilddrüse	♀ und ♂	29	0,29	0,15
Hyperthyreose	♀ und ♂	37	0,46	0,42
	♀	37	0,56	0,37
	♂	3	0,32	0,16
Schilddrüsencarcinom	♀	60	28	46
	♂	29	7,6	10

Anschriften: [1]) 8019 Dresden, Fetscherstr. 74; [2]) 8051 Dresden, PF 19

* Vortrag, gehalten auf 3. Tagung „Nukleare Analysenverfahren April 1983, Dresden

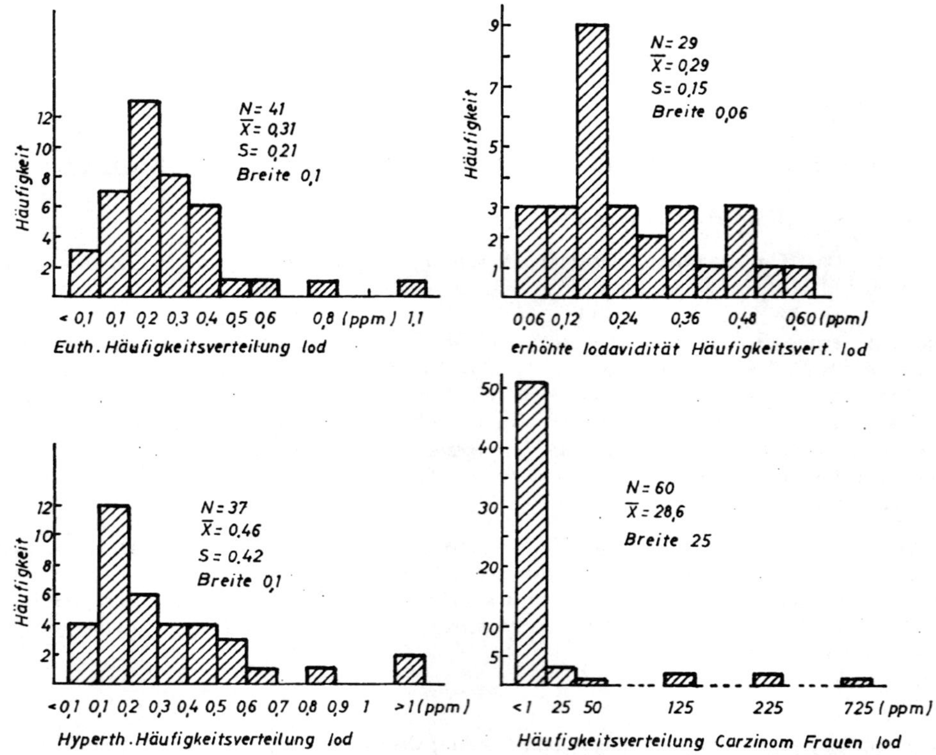

Abb. 1. Häufigkeitsverteilungen von Iodgehalten im Haar von 41 Patienten mit Euthyreose und Struma (links oben), 29 Patienten mit erhöhter Iodavidität der Schilddrüse (rechts oben), 37 Patienten mit Hyperthyreose (links unten) und 60 Patienten mit Schilddrüsencarcinom (rechts unten)

Iodavidität der Schilddrüse, mit Hyperthyreose kombiniert mit Struma und mit Schilddrüsencarcinom.

Der Iodgehalt im Haar von Patienten mit Schilddrüsencarcinom unterlag hohen individuellen Schwankungen. Die Mehrzahl dieser Patienten wies gegenüber der Kontrollgruppe mit Struma und Euthyreose bzw. Hyperthyreose einen erhöhten Iodwert im Haar auf. Für Frauen, es handelt sich um 66 Patientinnen, lag der Mittelwert bei 29 ppm, für Männer bei 7,6 ppm (29 Patienten). Die Ursache für diese Befunde konnte noch nicht endgültig geklärt werden. Es ist zu bedenken, daß in der Mehrzahl der Fälle eine Teilresektion der Schilddrüse oder eine Thyreoidektomie, d. h. eine operative Entfernung der Schilddrüse, der Entnahme der Haare vorausgegangen war. In Frage kommt aber auch ein direkter Zusammenhang mit dem Grundleiden, nämlich dem Schilddrüsencarcinom. Eine Iodexposition (z. B. Kontrastmittelgaben vor Haarentnahme) ging bei dieser Patientengruppe nicht mit einer generellen Steigerung der Iodgehalte im Haar einher. Die Häufigkeitsverteilung unterliegt nicht einer *Gauß*schen Verteilung. Abb. 1 unten rechts zeigt die Verteilung der Iodgehalte im Haar bei 60 Patientinnen mit Schilddrüsencarcinom.

Auf Grund unserer Untersuchungen scheint sich eine Differenzierung der Iodgehalte im Haar in Beziehung zum histologischen Befund des Schilddrüsencarcinoms anzudeuten. Bei Frauen betrug der Iodwert für follikulär-papilläre Ca 1,8 ppm, für papilläre Ca 28 ppm, für follikuläre Ca 36 ppm und für onkozytäre Ca 245 ppm. Diesen Befunden muß in Zukunft weiter nachgegangen werden, und sie sollten erneut beurteilt werden, sobald für jede Gruppe höhere Fallzahlen vorliegen. Möglicherweise spielen Aktivierungsgrad und Wachstumsrate eines Carcinoms eine größere Rolle für die Iodkonzentration im Haar als der histologische Befund.

Zur weiteren Präzisierung der Untersuchungen wurden die Konzentrationen der eingangs genannten Spurenelemente Zn, Se, Au, Cu, Mn, Na, Cl und Br in die Beurteilung einbezogen und ihre Korrelation zum Iodgehalt bestimmt. Es ergab sich, daß für Patienten mit Euthyreose eine relativ enge Korrelation des Iodgehaltes zum Gold- und zum Mangangehalt der Haare bestand. Der Korrelationskoeffizient belief sich für Iod/Gold auf 0,54 und für Iod/Kupfer auf 0,44. Für Patienten mit Schilddrüsencarcinom verringerten sich die Korrelationskoeffizienten auf 0,13 bzw. 0,11. Für die übrigen Spurenelementkonzentrationen bestand keine bzw. eine sehr schlechte Korrelation zum Iodgehalt für alle Patientengruppen. Aus diesen Untersuchungen ergibt sich die Schlußfolgerung, der Ursache der Korrelation zwischen Iod und Gold sowie Iod und Kupfer nachzugehen.

Tab. 2. Zn-Gehalt im Haar (ppm) von Patienten mit verschiedenen Schilddrüsenerkrankungen (Raum Dresden 1977–1982)

Gruppe		n	$\bar{x}$	s
Euthyreose und	♀	43	181	46
Struma	♂	3	188	41
Hyperthyreose und	♀	43	172	47
Struma	♂	–	–	–
Carcinom	♀	70	176	46
	♂	27	141	44

Mitteilungen
Information

ATOMIC WEIGHTS OF THE ELEMENTS 1981

The biennial review of atomic weight, determinations and other cognate data have resulted in the following changes in recommendes values (1979 values in parentheses): Hydrogen 1.00794 ± 7 (1.0079 + 1); Silver 107.8682 ± 3 (107.868 ± 1); Lutetium 174.967 ± 1 (174.967 ± 3). These values are incorporated in the Table of Standard Atomic Weights of the Elements 1981. Whereas in the past, the Table indicated uncertainties as either 1 or 3 in the last place, other single-digit uncertainties will in the future be quoted when there is convincing evidence that by their use, a more precise standard atomic weight can be tabulated. Important changes in annotations and the wording of footnotes to this Table and the Table of Isotopic Compositions are discussed. Changes in atomic weight values and in their estimated uncertainties in the period since 1969 are analysed. The Report includes for the first time a Table of Atomic Weights abbreviated to five significant figures in the expectation that changes in the tabulated data will rarely be needed. Attention is drawn to the possibility of materials being commercially available containing elements with unusual, atomic weights due to the enrichment or depletion of isotopes by free radical magnetic effects. The Relative Atomic Masses for Selected Radioisotopes are also tabulated. (N.E. Holden and R.L. Martin Pure and Appl. Chem. 55 (1983) 7, pp 1101-1118)

ISOTOPIC COMPOSITIONS OF THE ELEMENTS 1981

The Commission's biennial review of isotopic compositions as determined by mass spectrometry has been undertaken by the Subcommittee for the Assesment of Isotopic Composition (SAIC). A critical evaluation of the published literature forms the basis of the Table of Isotopic Compositions and Atomic Weights as Determined by Mass Spectrometry 1981, which is presented here. Atomic Weights calculated from the tabulated isotopic abundances are generally consistent with. Atomic weight values in the Table of Standard Atomic Weights 1981 with the most notable exceptions being the elements zinc, germanium and selenium, where the uncertainty intervals just meet or barely overlap. (N.E. Holden, R.L. Martin and J.L. Barnes, Pure and Appl. Chem. 55 (1983) 7, pp 1119-1136)

GEFAHREN UND VORTEILE VERSCHIEDENER ENERGIEVERSORGUNGSSYSTEME - SYMPOSIUMSBERICHT

Fachleute aus den verschiedensten Disziplinen, die zur Beurteilung der mit der Gewinnung und Verteilung von Energie verbundenen Gefahren und Vorteile herangezogen werden, trafen sich auf einem Symposium von IAEA/WHO/UNEP vom 9.-13.4.1984 in der KFA Juelich (BRD). Allen Referaten liegt die Erkenntnis zugrunde, dass keine Art der Energiegewinnung vollstaendig ohne Risiko sein kann. So werden beispielsweise bei der Verbrennung fossiler Brennstoffe unweigerlich Kohlendioxid sowie Schwefel- und Stickstoffoxide gebildet, die zur Entstehung des sauren Regens beitragen koennen. Bei der Befoerderung und Lagerung von verfluessigtem Erdgas und anderen fluessigen Kohlenwasserstoffen koennen sich Unfaelle mit ernsten Folgen ereignen. Bei Wasserkraftwerken hingegen besteht die Gefahr eines Dammbruches, wobei schon mehrfach das stromabwaerts gelegene Land verwuestet wurde und viele Opfer zu beklagen waren. Eine schwerwiegende Betriebsstoerung in einem Kernkraftwerk, die allerdings nicht sehr wahrscheinlich ist, kann die Bevoelkerung der Umgebung gefaehrden und das Krebsrisiko erhoehen. Selbst die sog. "alternativen" Energietraeger, wie Sonnen- und Windenergie, sind mit gewissen Gefahren verbunden, da die dabei benoetigten Materialen im Bergbaue gewonnen und betrachtet, die Anlagen errichtet und instandgehalten werden muessen, wobei es jederzeit zu Verletzungen oder toedlichen Unfaellen kommen kann. Bei der Wahl eines Energietraegers muessen nicht nur diese Risiken in Betracht gezogen werden, sondern auch die Vorteile aller vorhandenen Moeglichkeiten. Eine ausreichende Energieversorgung zu einem annehmbaren Preis ist massgebend fuer die wirtschaftliche Entwicklung eines Landes und damit auch fuer den Lebensstandard und die Beschaeftigungslage seiner Bevoelkerung. Bei der energiewirtschaftlichen Planung eines Staates muessen jedoch auch weltweit wirksame Faktoren beruecksichtigt werden, wie die vorhandenen Reserven, die Entwicklung der Brennstoffpreise auf dem Weltmarkt und die noetige Flexibilitaet fuer den Fall ploetzlicher Versorgungsengpaesse. Gleichzeitig muss man aber auch die Gefahren fuer die menschliche Gesundheit und die Umwelt in Betracht ziehen. Obwohl derartige Ueberlegungen die Entscheidung ueber die fuer ein Land geeignete Kombination von Energietraegern nicht sehr stark beeinflussen werden, muessen alle eventuellen Wirkungen auf ein Mindestmass herabgesetzt werden, wobei auch die Kosten zu beruecksichtigen sind. Es bedarf einer konsequenten Risikopolitik, um einen Ausgleich zwischen dem erwuenschten Ergebnis und den etwaigen negativen Folgen grossangelegter industrieller Unternehmen, wie der Energieproduktion, zu finden. (IAEA-Pressekommunique 84/10)

IAEA-NEUE NUKLEARDATEN

Der Anteil der Kernenergie an der Elektrizitaets-Erzeugung ist im Jahre 1983 erneut gewachsen. Das geht aus den Daten hervor, die von der IAEA gesammelt worden sind.
Der Netto-Zuwachs an Kernkraftwerken, die im Verlaufe des Jahres mit dem Elektrizitaetsnetz verbunden wurden, belief sich auf 23, so dass Ende 1983 insgesamt 317 Kernkraftwerke in 25 Laendern in Betrieb waren. Die Kernkraftwerks-Kapazitaet betrug damit ungefaehr 8 % der gesamten Elektrizitaets-Erzeugungs-Kapazitaet in der Welt. Die Kernkraftwerke erzeugten aber rund 12 Prozent der gesamten Elektrizitaet in der Welt, da Kernkraftwerke normalerweise im Grundlastbereich dauernd eingesetzt sind. Die Gesamtkapazitaet der 317 Kernkraftwerks-Einheiten belief sich auf 191 Gigawatt (GWe). Am Ende des Jahres 1983 waren zusaetzlich 209 Kernkraftwerke mit einer vorgesehenen Kapazitaet von 194 GWe im Bau.
Obwohl Kernenergie-Programme in einigen Laendern noch klein sind, uebersteigt der Anteil der Kernenergie an der Stromerzeugung in anderen Laendern zu Zeiten von geringer Nachfrage oft die 50-%-Marke. Die inoffizielle "Rangliste" fuer 1983, gestuetzt auf Daten des IAEA-Kernreaktor-Informations-Systems PRIS sieht folgendermassen aus:

Frankreich	48,3%	Japan	18,0%*
Belgien	45,9%	BRD	17,8%
Finnland	41,5%	Grossbritannien	17,0%
Schweden	36,9%	Kanada	12,9%
Taiwan, China	37,0%*	USA	12,6%
Bulgarien	32,3%	DDR	12,0%
Schweiz	29,3%	Ungarn	10,0%*
Suedkorea	18,4%		

(*geschaetzt)

Die IAEA nimmt an, dass die Gesamtkapazitaet an Kernenergie in der Welt bis zum Jahre 1985 auf 275 GWe ansteigen wird und zwischen 370 und 400 GWe im Jahre 1990 sowie zwischen 580 und 850 GWe am Ende unseres Jahrhunderts liegen wird. (IAEA-Pressekommunique 84/11)

STABLE ISOTOPE REFERENCE SAMPLES

"Advisory Group Meeting on Stable Isotope Reference Samples for Geochemical and Hydrological Investigations", IAEA, Vienna 19 - 21 September 1983, Report by R. Gonfiantini, March 1984, 77 pp.
Contents:
1. Introduction

2. Reference and intercalibration samples distributed by IAEA for stable isotope measurements in natural compounds,
   2.1 – V-SMOW (Vienna Standard Mean Ocean Water
   2.2 – SLAP (Standard Light Antarctic Precipitation
   2.3 – GISP (Greenland Ice Sheet Precipitation).
   2.4 – NBS-1 and NBS-1 A: water
   2.5 – NBS-19 and NBS-18: calcium carbonate
   2.6 – NBS-17 and NBS-16: carbon dioxide
   2.7 – NBS-30: biotite
   2.8 – OGS (Ocean Research Institute – Géological Survey of Japan Standard): barium sulphate
   2.9 – Soufre de Lacq: elemental sulphur
   2.10– IAEA-N-1 and IAEA-N-2: ammonium sulphate
3. Results of the stable isotope measurements intercomparison
   3.1 – Water samples (V-SMOW, SLAP, GISP)
      3.1.1 – Deuterium
      3.1.2 – Oxygen 18
   3.2 – Calcium carbonate samples (NBS-19, NBS-18)
      3.2.1 – Oxygen-18
      3.2.2 – Carbon-13
   3.3 – Carbon dioxide samples (NBS-17, NBS-16)
      3.3.1 – Oxygen-18
      3.3.2 – Carbon-13
   3.4 – Biotite sample (NBS-30)
   3.5 – Barium sulphate (OGS) and elemental sulphur (Soufre de Lacq)
      3.5.1 – Oxygen-18
      3.5.2 – Sulphur-34
   3.6 – Ammonium sulphate samples (IAEA-N-1 and IAEA-N-2)
4. Discussion, conclusions and recommendations
   4.1 – General remarks on the normalisation of isotope data
   4.2 – Hydrogen
   4.3 – Oxygen
   4.4 – Carbon
   4.5 – Carbon-13, deuterium and oxygen-18 in organic compounds
   4.6 – Sulphur
   4.7 – Nitrogen
5. Summary of the main conclusions and recommendations
   5.1 – Isotopic scales based on two samples with defined values
   5.2 – Re-evaluation of fractionation factors
   5.3 – New intercomparison samples
   5.4 – NBS samples to be sent to IAEA
   5.5 – Next meeting
   5.6 – Publication of a communication
   5.7 – Correspondance

References; 14 Table; 3 Figures; 3 Annexes

Kurznachrichten / Newlets

Isotope production:
PRODUCTION OF CARRIER-FREE Al-26

Al-26 is a nuclide of immense interest in the field of geochemistry. It is one of the primary nuclides used to determine the production rate of cosmic-ray-produced nuclides in meteorites and, by virtue of its long half-life ($7,5.10^5$y), to data those meteorites. For this use an accurate value of its half-life is extremely important. Other applications of Al-26 include studies of the mass transfer of Al^{3+} ions in liquid extractions and the use as a gamma-ray counting standard. For these reasons efforts were made to produce Al-26 through spallation reactions. A 63g electronics-grade single crystal silicon target was bombarded over a 2-year period with approximately 10^{23} protons having a nominal energy of 750 MeV. Al-26 was separated without adding Al-carrier using solvent extraction and ion-exchange methods. About $4,1.10^5$Bq or 560 µg of Al-26 were recovered along with $7,77.\ 10^{10}$Bq of Na-22 and $5,55\ .10^{10}$ Bq of Be-7 at the end of the bombardment. The tests were carried out at the Isotope Production Facility of the Los Alamos Meson Physics Facility. (Radiochimica Acta 33(1983) pp 213-215, Kimberly W. Thomas, The Los Alamos Medical Radioisotopes Group)

Metallorganische Verbindungen:
NESMEJANOW-GOLDMEDAILLE

Im Jahre 1982 wurde vom Praesidium der Akademie der Wissenschaften der UdSSR eine Goldmedaille ud ein Preis zu Ehren A. Nesmejanows, des hervorragenden sowjetischen Chemikers und Gruenders der wissenschaftlichen Chemie-Schule fuer metallorganische Verbindungen, gestiftet. Als erster erhielt diesen hohen Preis das Korrespondierende Mitglied der AdW der UdSSR M. Wolpin fuer eine Reihe von Forschungsarbeiten zur Synthese von neuen Typen metallorganischer Verbindungen und zur Untersuchung ihrer katalytischer Eigenschaften. M. Wolpin gelang es, in mehreren Reaktionen die Vorteile der neuentwickelten Katalysatoren gegenueber den herkoemmlichen zu beweisen. Die Verbindungen von Graphit und Uebergangsmetallen wirken als Katalysatoren bei einer so einmaligen Reaktion wie der Umwandlung von Graphit in Diamanten, d.h. diese Arbeit legte den Grundstein fuer einen prinzipiell neuen Bildungsmechanismus sowie fuer ein eigenstaendiges Herstellungsverfahren von synthetischen Diamanten. Die Verbindungen von Graphit und Uebergangsmetallen sind imstande, graphitaehnliche Strukturen umzubauen. So liegt es nahe, diese Katalysatoren auch bei der Umwandlung von Kohle in Kohlenwasserstoffe einzusetzen. (Wissenschaft in der UdSSR (1984) 1)

Strahlenschutz:
HOHE RADON-BELASTUNG

Die Eidg. Kommission zur Ueberwachung der Radioaktivitaet (KUER) hat in ihrem Bericht fuer das Jahr 1982 erstmals die Ergebnisse einer Stichprobenerhebung ueber Rn in der Atemluft in der Schweiz aufgefuehrt. Das natuerliche radioaktive Edelgas Rn stammt vor allem aus dem Erdboden sowie aus Baustoffen. Untersuchungen in verschiedenen Laendern haben ergeben, dass durch Massnahmen zur Abdichtung der Haeuser aus Gruenden der Energieeinsparung ein erheblicher Anstieg der Rn-Konzentration in den Wohnraeumen eintritt, falls nicht die Ventilationsrate erhoeht wird. Nach dem Bericht der KUER erhaelt der Durchschnittsschweizer pro Jahr eine Strahlendosis von ca. 400 mrem. Sie setzt sich zusammen aus 125 mrem durch natuerliche Strahlenquellen, 150 mrem durch Rn in Haeusern, 120 mrem durch Roentgenuntersuchungen und weniger als 10 mrem durch weitere kuenstliche Strahlenquellen. In diesen 10 mrem pro Jahr ist der Beitrag der Kernkraftwerke enthalten, die den Schweizer im Durchschnitt mit weit unter 1 mrem pro Jahr belasten, ein Wert, wie er auch in den alljaehrlichen Berichten ueber die Strahlenbelastung in der BRD ausgewiesen wird. (Atomwirtschaft/Atomtechnik 29 (1984) 3, S. K+U IV)

Kernenergie:
STATISTISCHES ZAHLENSPIEL

Im "Bulletin SEV/VSE" bringt der Verband Schweizerischer Elektrizitaetswerke Beispiele fuer statistische Zahlenspielereien. Die Schweiz weist fuer das Jahr 1982 einen Strom-Endverbrauch von 36,7 Terawattstunden (TWh) oder 0,13Exajoule (EJ) aus. Ist das viel – ist das wenig? Dies entspricht 36 700 Mill. Kilowattstunden, was schon nach sehr viel klingt. Und falls die offiziellen internationalen SI-Einheiten Joule (J) verwendet werden, so ergibt sich eine fast astronomische Zahl von 132 Billionen. An der Eidg. Techn. Hochschule in Zuerich betraegt der jaehrliche Stromverbrauch pro Student rund 9000 kWh, wobei der Chemiestudent mit 24000 kWh die Spitze haelt. Rechnet man die kWh zu zwoelf Rappen, so verbraucht jeder ETH-Student im Mittel fuer ueber 1000 Franken Strom. Rund das Doppelte seiner jaehrlichen Studiengebuehren. Ein Beschaeftigter in der Schachtelkaesebranche verbraucht im Jahresdurchschnitt rund 6000 kWh – eine Milchvieheinheit (Melken, Kuehlung, Futtereinlagerung, Heubelueftung) etwa 650 kWh. Statistisch betrachtet entsprechen einem ETH-Studenten rund 14 Milchvieheinheiten. (Atomwirtschaft/Atomtechnik 29 (1984) 3, S. K+U IV)

Mass Spectrometry:
ISOTOPE RATIO OF N_2 AND O_2

A cryogenic separation system has been developed for the quantitative separation of nitrogen and oxygen for high-precision mass spectrometric isotopic determinations. The system is unique in that it does not require conversion of oxygen to CO_2 which allows determination of the $^{17}O/^{16}O$ ratio. Nitrogen and oxygen isotopic ratios obtained are identical with those previously reported for air values, i.e. nitrogen shows that, in each case, they are isotopically identical with the atmospheric ratio $\delta^{15}N=0$ within the mass spetrometric error and for oxygen $\delta^{18}O$ = (23,50 ± 0,3)‰ in a range extending from 23,18‰ .... 23,79‰. The ^{17}O value is, within error, identical with that expected from $\delta^{18}O$, i.e. $\delta^{17}O = 0{,}52 \cdot \delta^{18}O$. (Cryogenic separation of N_2 and O_2 in air for determination of isotopic ratios by mass spectrometry, M.H. Thiemens and D. Meagher, Analyt. Chem. 56 (1984) 2, pp 201–203)

Veröffentlichungen / Publications

UNITED KINGDOM ATOMIC ENERGY AUTHORITY

AERE-R 10859
Radioactive fallout in air and rain: results to the end of 1982 – Ju e, 1983, 40 pp. R.S. Cambray et al.

AERE-R 10885
Ambient ozone measurements at Harwell, June-December 1982 – June, 1983, 27pp. F.J. Sandalls et al.

AERE-R 10939
Biological and chemical characterisation of domestic refuse landfills (ii)L-field landfill... – June, 1983, 16pp. J.F. Rees et al.

AERE-M 3350
Differential neutron cross section of ^{47,48}Ti (n,p) ^{47,48}Sc and ^{27}Al(n, α)^{24}Na – September, 1983, 16pp. S.J. Firkin

AERE-M 3365
Inelastic neutron scattering and fission in U^{238} – November, 1982, 8pp. D. Wilmore

AERE-R 10787
Non-intrusive measurements and numerical comparison of the axial velocity components in a two-dimensional flow channel for a backward facing step and a rib roughened surface – May, 1983, 33pp. C.J. Bates et al.

AERE-R 10895
Radiation-induced segregation in Fe-Cr-Ni alloys – May, 1983, 16pp. A.D. Marwick et al.

AERE-R 10907
An analysis of grain boundary failure in austenitic stainless steel – May, 1983, 19pp. K.L. Kear et al.

AERE-R 10795
The density effect for fast ions – September, 1983, 123pp. C.J. Woods

AERE-R 10882
NAMSOL: Finite elements program for migration of radionuclides in groundwater – September, 1983, 72pp. E.A. Dolman et al.

AERE-R 10943
The removal of strontium activity from magnesium containing effluents by barium sulphate precipitation and ultrafiltration – August, 1983, 28pp. R.G. Gutman et al.

AERE-R 11029
The application of HTR coated fuel particle performance models to various problems in materials' science – October, 1983, 37pp. D.G. Martin

Veranstaltungskalender – Tagungen – Lehrgänge / Calendar – Meetings – Courses

KURZVERANSTALTUNG "TECHNOLOGIE UND WIRTSCHAFT – DER EUROPAEISCHE BEITRAG"

Veranstalter: Forschungszentrum Seibersdorf gemeinsam mit dem Oesterreichischen College und der Bundeskammer der gewerblichen Wirtschaft im Rahmen des 40. Europaeischen Forums, Alpbach/Oesterreich, am 20./21. August 1984. Weitere Auskuenfte: Dr. G. Hillebrand und Dipl.- Ing. H. Gutmann, Oesterreichisches Forschungszentrum Seibersdorf GmbH.

Ausgewählte Beiträge / Recently published papers

ATOMNAJA ENERGIJA 57(1984)1 (all in Russian)

On the Introduction of the International System of Units for Ionizing Radiation – V.V. Bochkaryov et al.

Accumulation of Gaseous Products from Nuclear Reactions in Chromium and Nickel During Irradiation by High-Energy Electrons – A.G. Zaluzhny et al.

Techniques for Corrosion Testing of Zirconium Alloys in a Research Reactor – B.V. Samsonov et al.

Effect of Aluminum, Beryllium and Chromium (III) Nitrates on Uranium Nitrate Extraction by 10% Butyl Phosphate Solution in n-Alkanes – B.S. Zakharkin et al.

On the Creation of Automated Environmental Radiation Monitoring Systems at Atomic Power Plants – E.P. Volkov et al.

Background Radiation in Dwellings – E.M. Krisyuk et al.

Changes in the Mechanical Properties of Loaded Steel Samples when Electrolytically Saturated with Hydrogen – J. Bidzhero et al.

Economic Aspects of Using Bremsstrahlung in Radiation Technology – V.V. Krayushkin et al.

The Diffusion of Fission Products Ru-103 and Zr-95 in Monocrystallic Tungsten – D.K. Daukeyev et al.

Anomaly Effects of Small Doses of Ionizing Radiation on Metals and Alloys – I.P. Chernov et al.

The Measurement of Cm-244 Cross-Section by Fast Neutrons Using a Nanogram Quantity of Isotope – P.E. Vorotnikov et al.

GEOL. SOC. AMER. BULL. 95(1984)1

Alluvial stratigraphy and radiocarbon dating along the Duck river, Tennessee – Implications regarding flood-plain origin – G.R. Brakenbridge

ECONOMIC GEOLOGY 79(1984)1

Textural and stable isotope studies of the Big Mike Cupuferrous volcanogenic massive sulfide deposit, Persking County, Nevada – R.O. Ry et al.

GEOCHEMICAL J. 17(1983)6

K-Ar ages of dikes in Northeast Japan – H. Tsunakawa et al.

Compretensive Sr isotope study of Quaternary volcanic rocks from the Izu-Ogasawora arc – K. Notsu et al.

Apporent penetration of fallout C-137 into rocks in Izu-Oshima – K. Kimura et al.

Be-10 and Be-10/Be-9 in manganese nodules – T. Inoue et al.

An examination of geochronological utility of electron capture decay of La-138 – A. Masuda et al.

EARTH PLANET SCI. LAKES 67(1984)1

C-14 crulent of 10 meteorites by tandem accelerator mass spectrometry – R.M. Brown et al.

The growth of the continent through geological time studied by Nd isotope analysis of shales – C.J. Allegre et al.

J. CHEM. PHYSICS 80(1984)1

Fluorescence of excited ethynyl radicals produced by pulsed vacuum ultraviolet photolyses of C_2H_2, C_2D_2, and C_2HBr –Y. Saito et al.

Raman spectroscopic study of $EtNH_3X$ (X=Cl,Br) and several deuterated analogs – H. Hagemann et al.

Mass spectrometric determination of partial and total electron impact ionization cross sections of SO_2 from threshold up to 200 eV – O.J. Orient et al.

Absolute rate constans for the reaction of $O(^3P)$ atoms with ethylene, propylene, and propylene-d6 over the temperature range 258-861 K – R.A. Perry

Dissociation of vibronic states of $C-^1A'$ DCN: Quantum treatment – D.T. Chuljian et al.

Dependence of reaction attributes, Including differential cross sections and resonance features, on changes in the potential energy surface for the $F+D_2$ reaction – R.B. Walker et al.

Classical barrier height for $H+H_2 \longrightarrow H_2+H$ – B. Liu

Thermal diffusion factors for the system Ar-Kr at 300 and 255 K – R.D. Trengove et al.

Deuterium isotope effects on the nitrogen chemical shift and $^1J(N,H)$ in the ammonium ion – R. E. Wasylishen et al.

J. CHEM. PHYSICS 80(1984)2

Electronic excited states of CO_2: An electron impact investigation – R. McDiarmid et al.

Trajectory studies of model $H-C-C \longrightarrow H+C=C$ dissociation. III. Details of the lifetime distribution following chemical activation – W.L. Hase et al.

Absolute production rates and efficiencies of NO in electron-irradiated N_2/O_2 mixtures – B.D. Green et al.

ND_4 Schueler band absorption observed by laser FM spectroscopy in a photochemical reaction – E.A. Whittaker et al.

CHEMICAL PHYSICS LETTERS 103(1984)5

The magnetic isotope effect and oxygen isotope selection in chain processes of polymer oxidation – A.L. Buchachenko et al.

GEOKHIMIJA (1983)Nov. (all in Russian)

A.Je. Fersman (100 years birthday jubelee) – G.B. Naumov

Element and isotope ratios of juvenile noble gases – A.B. Verkhovsky et al.

Radiogeochemical investigation of the copper-porphyric ore deposits and the possibility of its radiometric prospecting – N.G. Syromyatnikov et al.

J. CHEM. PHYSICS 79(1983)12

The microwave spectra of CH_2DSH and CHD_2SH – Chun Fu Su et al.

Separation of O, C, and S isotopes by terostep laser photodissociation of OCS – P.F. Zikel et al.

The measurement of diffusion using deuterium pulsed field gradient NMR – P.T. Callaghan et al.

Determination of the spectrum of recoil energies for chlorine atoms generated by the Cl-35 (n,γ) Cl-36 nuclear process – J. Chang et al.

ANGEW. CHEM. 96(1984)1

Destabilisierte Carbokationen – ein Ueberblick (109 Zitate) – Th.T. Tidwell
(eine elegante Technik zur Erzeugung von freien Methylkationen in Loesung, beruht auf dem Zerfall von $CT_4 \longrightarrow CT_3^+ + {}^3He + \beta^-$)

Ein neuer aromatischer Kohlenwasserstoff mit 16 -Elektronen-Peripherie : "Cyclisches Bicalicen" – Shigeo Yoneda et al.

Zum Mechanismus der Cyclopropan- "walk"-Umlagerung : Synthese und Eigenschaften des 2-Diazopropan-Addukts von Cyclooctotetraen – F.-G. Klaerner et al.
(Klaerung durch Deuterium-Markierung)

Hydroxidion-initiierte Deuterierung sehr schwach (H-acider Verbindungen nach Phasentransfer-Katalyse-Bedingungen – M. Halpern et al.

ATOMKERNENERGIE-KERNTECHNIK 44(1984)2

Wo stehen wir heute bei der Realisierung des Endlagerkonzeptes – E. Merz

Ableitung von Anforderungen an endzulagernde Abfallgebinde und deren Pruefung – G.G. Eigenwillig u.a.

Behandlung von Spaltproduktloesungen mittlerer und hoher Radioaktivitaet – R. Odoj u.a.

Behandlung und Beseitigung der fluechtigen Radionuklide Tritium, Kohlenstoff-14, Krypton-85 und Jod-129 bei der Wiederaufarbeitung – H. Bruecher

Direkte Endlagerung abgebrannter Brennelemente – K.-D. Closs

Abfallbehandlung bei Stillegung und Beseitigung kerntechnischer Anlagen – H. Hepp

Strahlenschutz und Radiooekologie der Endlagerung – H. Illi u.a.

Sichere Lagerung von hochradioaktiven verglasten Abfaellen im Uebergangslager – P.-W. Phlippen u.a.

Moegliche Anwendungen der Kernspurtechnik – B.E. Fischer u.a.

Untersuchungen von Spurenaetzparametern in Festkoerper-Kernspurdetektoren – H.A. Khan

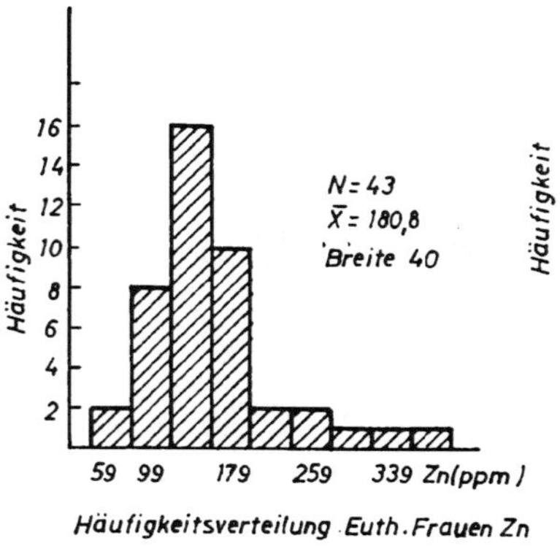

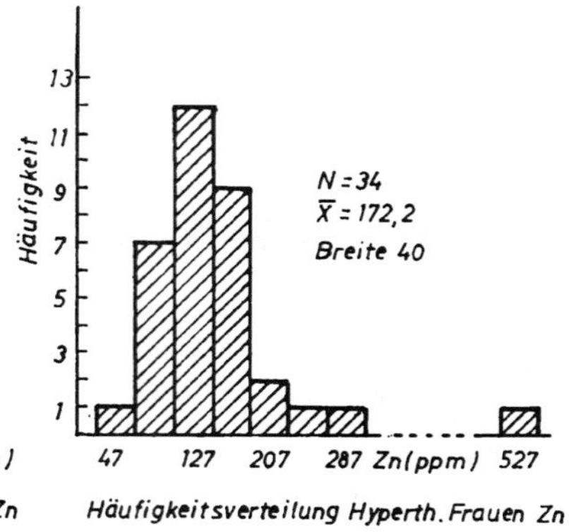

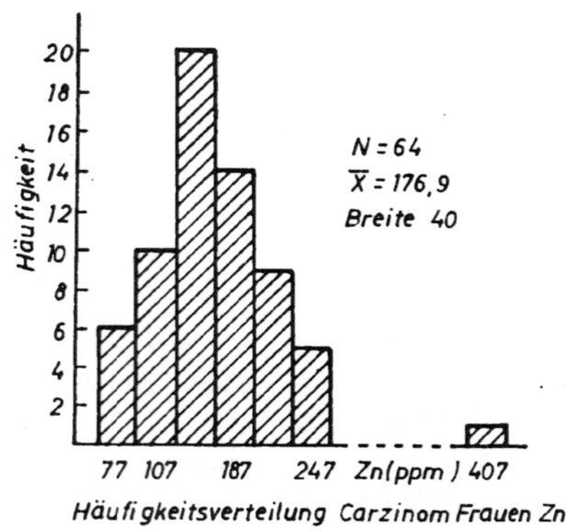

Abb. 2. Häufigkeitsverteilungen von Zinkgehalten im Haar von 43 Patientinnen mit Euthyreose und Struma (links oben), 34 Patientinnen mit Hypterthyreose (rechts oben) und 65 Patientinnen mit Schilddrüsencarcinom (rechts unten)

Korrektur: Statt $N = 64$ lies $N = 65$

Ein weiterer Schwerpunkt unserer Untersuchungen lag auf der Bestimmung der Konzentration von Zink bei verschiedenen Schilddrüsenerkrankungen (Tab. 2). Die Zinkkonzentration im Haar wies im Gegensatz zu den anderen ebenfalls geprüften Spurenelementkonzentrationen bei Patientinnen mit Euthyreose, Hyperthyreose und Carcinom etwa die gleiche Häufigkeitsverteilung auf. Abb. 2 verdeutlicht dieses Verhalten für Patientinnen mit Euthyreose und Struma, Hyperthyreose und mit Schilddrüsencarcinom. Da auch die Mittelwerte der Zinkkonzentrationen im Haar für die genannten Krankheitsbilder keine Unterschiede aufweisen, ist für unser Patientengut, d. h. für Schilddrüsencarcinomträger, im Gegensatz zu den bekannten Tatsachen bei der Diagnostik des Mammacarcinoms [1] nicht mit Veränderungen der Zinkkonzentration im Haar zu rechnen, die eine Differenzierung von Schilddrüsenerkrankungen erlauben. Im Vergleich zur Kontrollgruppe ist bei den weiblichen Strumaträgern allerdings eine leichte Erhöhung der Zinkgehalte im Haar zu beobachten.

Aus unseren Untersuchungen ergeben sich folgende Schlußfolgerungen:

- Die klar definierten Funktionslagen von Strumaträgern erweisen sich hinsichtlich des Iodgehaltes im Haar als relativ homogene Gruppen.
- Strumaträger weisen erniedrigte Iodwerte im Haar auf. Diese Erkenntnis fügt sich ein in die Theorie über die Entstehung der Struma, bei der Iodmangel diskutiert wird.
- Der Iodgehalt im Haar bei Hyperthyreose ist erhöht im Vergleich zu Strumaträgern mit Euthyreose.
- Die Mehrzahl der bei uns zur Untersuchung gekommenen Patienten mit Schilddrüsencarcinomen weisen stark erhöhte Iodkonzentrationen im Haar auf. Diese Tatsache sollte bei der Diagnostik von Schilddrüsencarcinomen in Betracht gezogen werden. Evtl. bietet sich hier eine Möglichkeit der Verbesserung der Frühdiagnostik und Verlaufskontrolle von Schilddrüsencarcinomen an.
- Iodbestimmungen im Haar eignen sich zur Beurteilung der Iodreservoirs in Geweben und ermöglichen damit eine neue diagnostische Aussage, die bisher noch nicht getroffen werden konnte.
- Die Iodbestimmung im Haar ist für epidemiologische Untersuchungen geeignet.
- Die Einbeziehung weiterer Spurenelementkonzentrationsbestimmungen in die Untersuchungen, insbesondere von Kupfer und Gold, sowie die Bestimmung des Verhältnisses zum Iodgehalt ist zur Präzisierung der Diagnostik von Schilddrüsenerkrankungen zu empfehlen.

Eingegangen am 12. 10. 1983

Literatur

[1] *K. Baldauf, B. Schicke, W. Wiesener*, Arbeitstagung Mengen- und Spurenelemente (1982) 475

[2] *G. V. Iyengar, W. E. Kolmer, H. J. M. Bowen*, The Elemental Composition of Human Tissues and Body Fluids, Weinheim – New York (1978)

[3] *T. Mende, W. Wiesener, W. Görner*, Zbl. Pharm. **121** (1982) 427

[4] *W. Wiesener, W. Görner, S. Niese, K. Baldauf, W. Grund, M. Hennig, T. Mende*, Isotopenpraxis 17 (1981) 278

[5] *Yu. S. Ryabukhin*, Report IAEA/RL 50 (Okt. 1978)

Isotopenpraxis 20 (1984) 8, S. 304 bis 305

Effect of Epinephrine and Insulin on Glucose Metabolism in Isolated Fat Cells

M. A. Zewail

(National Research Centre, Dokki, Cairo)

N. M. Fahmy

(National Organisation for Drug Control and Research, Cairo, Egypt)

J. H. Nielsen

(Niels Steensen Hospital, Gentofte, Denmark)

Since epinephrine and insulin are known to counteract each other in several metabolic respects, it was found of interest to evaluate the interaction of these hormones on glucose metabolism in isolated fat cells. Results indicate that in spite of reduced lipolysis, insulin potentiates the conversion of glucose to glyceride-glycerol. This means that a certain degree of esterification of fatty acids must take place. So the low rate of lipolysis is sufficient to provide fatty acids for this esterification.

Da bekannt ist, daß Epinephrin und Insulin in mehrfacher metabolischer Hinsicht einander entgegenwirken, war es interessant, die Wechselwirkung dieser Hormone hinsichtlich des Glukosemetabolismus in isolierten Fettzellen zu ermitteln. Die Ergebnisse weisen darauf hin, daß das Insulin trotz reduzierter Lipolyse die Umwandlung von Glukose in Glyzerid-Glyzerol ermöglicht. Das heißt, daß in einem bestimmten Umfang eine Veresterung von Fettsäuren erfolgen muß. Daher ist die niedrige Lipolyserate ausreichend, um die Fettsäuren für diese Veresterung bereitzustellen.

Keywords

adrenaline; carbon 14 compounds; glucagon; glucose; receptors

Introduction

In earlier experiments [11] it was found that glucagen depressed the conversion of glucose to glyceride fatty acids and stimulated the synthesis of glyceride-glycerol.

In the presence of insulin, glucagon did not influence the increased fatty acid synthesis, while the glycerol formation was potentiated at high glucagon concentrations.

This effect seemed not to depend on the lipolytic effect of glucagon. Since epinephrine resembles glucagon in its lipolytic action by stimulating the cyclic-AMP formation via B-adrenergic receptors [8–10], but differs from glucagon in stimulating lipolysis in cells from old heavy rats and in its action on adrenergic receptors, the present study was performed.

Materials and Methods

Animals

The experimental animals consisted of Wistar male rats weighing 200 g and aged 60 days or weighing 320 g and aged 90 days. They were fed ad libitum with standard laboratory chow, Rostock pellets, which contained 21% protein, 56% carbohydrate and 23% fat.

Preparation and Incubation of Isolated Fat Cells

The animals were sacrificed by decapitation on the same day of the experiment and the fat pads were removed and used for the preparation of isolated fat cells according to [3].

The free fat cells were suspended under gentle magnetic stirring in HEPES buffer solution of pH 7·4 with the addition of glucose 0·1 mg per ml and U-(^{14}C)-glucose, 1 μCi/ml. 1 ml of cell suspension was incubated with insulin and/or epinephrine at the appropriate concentration. CO_2 was collected as described by *Gliemann* [2].

Extraction of triglyceride was performed with heptane in isopropanol according to [2, 3, 7]. The lipolysis was measured by suspending the fat cells in buffer solution without glucose, and placing 1 ml in tubes with the appropriate hormones.

After incubation in a metabolic shaker for 1 hour at 37 °C, the incubation was stopped by the addition of 1 ml 10% (w/v) trichloroacetic acid. Glycerol release was determined as described by *Korn* [6]. Cell numbers were determined under the microscope after fixation with osmium-tetroxide.

Results

Fig. 1 shows the lipolysis in absence (A) and presence of glucose 1 mg per ml (B) No qualitative difference but a

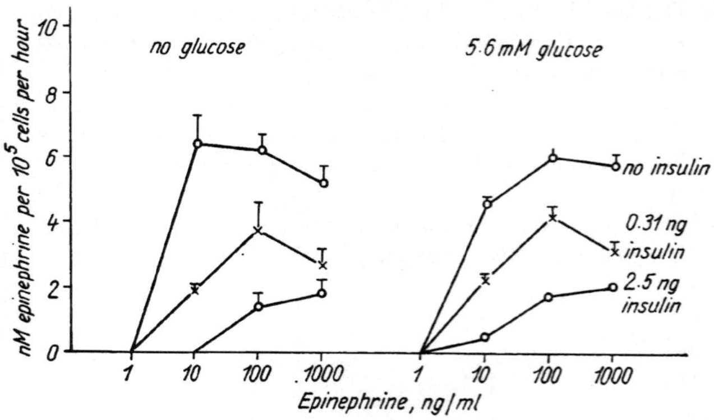

Fig. 1. Effect of epinephrine and insulin on glycerol release in absence and presence of glucose

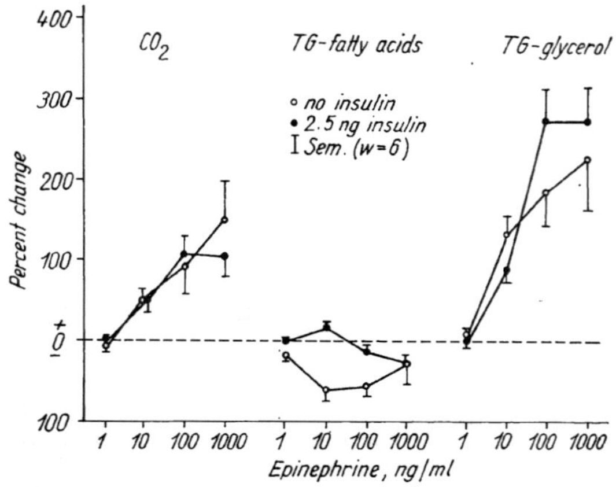

Fig. 2. Effect of epinephrine on glucose metabolism in fat cells from male rats (123 – 292 g)

small quantitative difference is seen between the effect of epinephrine and insulin in the two situations.

Fig. 2 shows the effect of epinephrine and insulin on glucose metabolism in fat cells from rats weighing below 300 grams. In the absence of insulin, epinephrine is seen to decrease the rate of fatty acid synthesis. In the presence of insulin, potentiation of glycerol synthesis was seen.

The potentiation was lower at low epinephrine concentration, but markedly increased at higher epinephrine concentrations. In the experiments with fat cells from old heavy rats weighing more than 300 grams, a potentiation of the glycerol synthesis was also seen.

Discussion

The stimulation effect of epinephrine on glycogenolysis and lipolysis is attributed to its ability to elevate levels of cyclic-AMP in appropriate tissues, while the antagonistic effect of insulin in this respect may be due to its ability to lower levels of cyclic-AMP in several tissues [1, 4, 5].

The results of the present study indicate that in spite of reduced lipolysis, insulin potentiates the conversion of glucose to glyceride-glycerol. This means that a certain degree of esterification of fatty acids must take place. So the low rate of lipolysis is sufficient to provide fatty acids for this esterification. However, uptake of albumin bound fatty acids from the medium cannot be excluded. The reason for the increased labelling of the glycerol could be explained from a decrease in glucose supply from preexisting glycogen stores in the fat cells. The same phenomenon was seen with glucagon, i. e. at low concentrations of the lipolytic hormone insulin can suppress the lipolysis and new synthesis of glycerol. While at higher concentrations of lipolytic hormone, the lipolysis will provide fatty acids for esterification of newly synthesized glycerol-phosphate mainly originating from external glucose which may be more pronounced because of reduced supply of endogeneous glucose from glucogen.

Received June 2, 1983
Accepted im revised form November 9, 1983

References

[1] *R. W. Butcher, J. G. T. Sneyd, C. R. Park, E. W. Sutherland,* J. Biol. Chem. **241** (1966) 1651
[2] *J. Gliemann,* Diabetologia **3** (1967) 382
[3] *F. M. Hansen, J. H. Nielsen, J. Gliemann,* Eur. J. Clin. Invest. **4.** (1974) 411
[4] *R. L. Jangas,* Proc. Nat. Acad. Sci. U.S.A. **56** (1966) 757
[5] *L. S. Jefferson, J. H. Exton, R. W. Butcher, E. W. Sutherland.* J. Biol. Chem. **243** (1968) 1031
[6] *E. D. Korn,* J. Biol. Chem. **251** (1955) 1
[7] *M. Rodbell,* J. Biol. Chem. **239** (1964) 375
[8] *M. A. Rizak,* J. Biol. Chem. **239** (1964) 392
[9] *G. A. Robinson, R. W. Butcher, E. W. Sutherland,* Ann. N. Y. Acad. Sci. **139** (1967) 703
[10] *G. A. Robinson, R. W. Butcher, E. W. Sutherland,* Ann. Rev. Biochem. **37** (1968) 149
[11] *M. A. Zewail, J. H. Nielsen,* FEBS MEETING, Copenhagen, Denmark 1977

Isotopenpraxis **20** (1984) 8, S. 306 bis 309

Hormonal Interaction in Diabetic Pregnancy

A.-R. A. Hafiez[1]), *M. A. Abdel-Hafez*[1]) *E. A. Osman*[2]), *M. S. Ibrahim*[3])

([1]) Biochemistry Dept.; [2]) Gynaecology and Obstetrics Dept., Kasr El-Aini Faculty of Medicine, Cairo University; [3]) Internal Medicine Dept. Al-Azhar University)

Serum glucose, human placental lactogen (HPL), prolactin (PRL), estradiol (E_2), progesterone (P), cortisol and human growth hormone (HGH) were determined in nondiabetic (19 cases) and diabetic (19 cases) pregnant women during the 32nd and 36th week of gestation. Significant elevation of HPL, PRL, HGH and cortisol was found in the diabetic pregnant women during the 32nd week while E_2 and P were not significantly changed from the corresponding levels in the nondiabetic group. One can conclude that the changes in the hormonal pattern during gestation may induce carbohydryte intolerance observed in diabetic pregnancies.

Bei 19 nichtdiabetischen und bei 19 diabetischen Schwangeren wurden in der 32. und in der 36. Woche der Schwangerschaft Serumglukose, Humanplazentalaktogen (HPL), Prolaktin (PRL), Estradiol (E_2), Progesteron (P), Cortisol und humane Wachstumshormone ermittelt. Bei diabetischen Schwangeren in der 32. Woche wurde eine wesentliche Erhöhung bei HPL, PRL, HGH und Cortisol festgestellt, während die E_2- und P-Werte gegenüber den entsprechenden Werten der nichtdiabetischen Gruppe keine wesentlichen Veränderungen aufwiesen. Man kann daraus schließen, daß die Veränderungen im Hormonhaushalt während der Schwangerschaft Einfluß auf die bei diabetischen Schwangeren beobachtete Kohlenhydrat-Unverträglichkeit haben kann.

Keywords

cortisone; diabetes mellitus; endocrine diseases; HPL; pregnancy; progesterone; prolactin; steroid hormones

Introduction

During normal pregnancy certain metabolic changes tend to disturb the homeostasis of carbohydrate metabolism with consequent lowring of glucose tolerance. These changes might alter the euglycaemic state for the nutritional requirements of the fetus.

Placental hormones are accused of alteration of metabolic pathways. Among these hormones is the human placental lactogen (HPL) which has been subject of great disagreement. Some authors classify it as diabetogenic [1] while others deny this [2]. The role of prolactin remains to be elucidated. The aim of this work is to find-out any correlation between serum HPL, PRL, E_2, progesterone, human growth hormone (HGH), and cortisol in pregnant nondiabetic and diabetic women during the same period of gestation.

Material and Methods

Thirty-eight pregnant women in the 32nd week of gestation were selected for the study (nineteen were diabetics and nineteen were nondiabetic). Eleven of the latter and nine of the former were re-investigated during the 36th week. All the diabetics were insulin dependent, but recieved no insulin on the day of the test. The following biochemical parameters were assayed: serum glucose level by glucose oxidase method [3], HPL [4], E_2 and progesterone [5], HGH [6], PRL [7], and cortisol [8]. Hormonal assays were done by double antibody technique of RIA using kits provided by Immunochem Corp. Calif. Two-hours after 100 g oral glucose load another venous blood samples were taken and serum levels of glucose and HPL were reestimated. Results were statistically analysed according to the student's "t" test.

All patients were examined thoroughly to exclude superimposed toxaemia. They were observed after delivery to evaluate the fetal condition as regards the weight and Apgar scores.

Results

The results obtained are given in Tab. 1—3 and Fig. 1 and 2. Tab. 1 and Fig. 1 showed changes in serum glucose levels at fasting and 2 hours after glucose meal in the nondiabetic and diabetic groups during the 32nd week of gestation. Glucose intolerance was markedly altered in both groups. The criteria was repreated during the 36th week

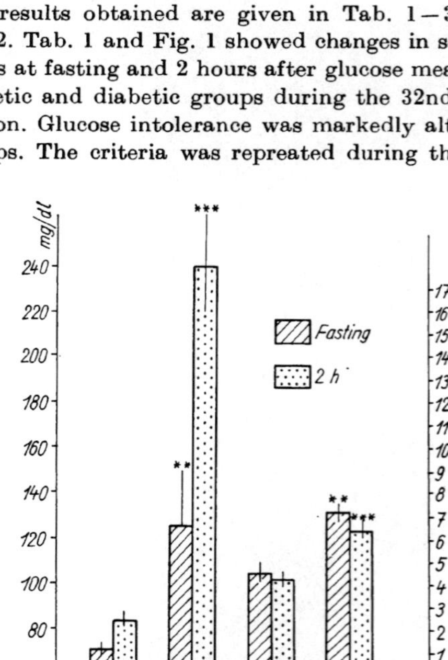

Fig. 1. Serum glucose (mg/dl) and HPL (μg/ml) in non-diabetic and diabetic pregnant women during the 32nd week

Tab. 1. Serum glucose changes (mg/dl) in non-diabetic and diabetic pregnant women at 32nd and 36th weeks

Time		Non diabetics		Diabetics	
		fasting	2 hours	fasting	2 hours
	M	72	84	128	236
32^{nd}w	±SE	2·33	2·69	18·78	20·83
	P_1			0.01	0·001
	M	70	78	117	239
36^{th}w	±SE	2·92	6·29	10·6	23
	P_1			0·005	0·001
	P_2	NS	NS	NS	NS

P_1 = Serum glucose of the diabetics vs the corresponding nondiabetics.
P_2 = Serum glucose in 36th week vs the corresponding 32nd week levels.

Tab. 2. Fasting serum hormone levels in non-diabetic and diabetic pregnancies at 32nd week of gestation

		HPL µg/ml	PRL ng/ml	HGH ng/ml	Cortisol µg/dl	E_2 µg/ml	Prog. ng/ml
Non Diabetic	M	4·5	252	1·2	18	10·8	76
(19)	±SE	0·3	23·7	0·08	1·48	1·4	12·9
	M	7·1	404	1·8	25	12·1	95
Diabetics (19)	±SE	0·46	41·8	0·3	2·9	1·26	12·9
	P	0·01	0·005	NS	0·05	NS	NS

Tab. 3. Fasting serum hormone levels in non-diabetic and diabetic pregnancies at 36th week of gestation

		HPL µg/ml	PRL ng/ml	HGH ng/ml	Cortisol µg/dl	E_2 µg/ml	Prog. ng/ml
Non Diabetic	M	4·6	338	1·2	21	26	85
(11)	±SE	0·3	41·1	0·1	2·9	10·8	16·2
	P_1	NS	NS	NS	NS	NS	NS
Diabetic	M	8·0	558	1·8	28	45	135
(9)	±SE	4·4	76·5	0·3	1·6	12·3	29·8
	P_1	0·001	0·025	NS	0·05	NS	NS
	P_2	NS	NS	NS	NS	0·01	NS

P_1 = Results of Non diabetics at 32nd w. vs 36th w.
P_2 = Diabetics vs the non diabetics during 36th w.

of gestation Tab. 1. No significant changes in the serum glucose levels during the 32nd and 36th weeks (P was NS).

Human placental lactogen (HPL) at "F" hour was significantly higher in the diabetic pregnant women (Tab. 2) $P < 0{\cdot}01$. No significant elevation was observed after ingestion of glucose in both nondiabetic and diabetic groups.

Fasting serum levels of PRL, Cortisol, were significantly higher in the diabetic women than the corresponding levels in nondiabetic group (Tab. 2 and Fig. 2). Nonsignificant rise in HGH was observed. E_2 and progesterone levels were not altered significantly.

On the 36th week of gestation no significant changes were observed in both groups in the levels of all hormones.

Discussion

The presented data showed clearly that a change in the hormonal patterns does exist in nondiabetic and diabetic pregnancies in the two gestational periods studied (32nd and 36th weeks). The intolerance of glucose observed in the diabetic group was associated with significant elevation in the levels of serum HPL, PRL, HGH and cortisol. This

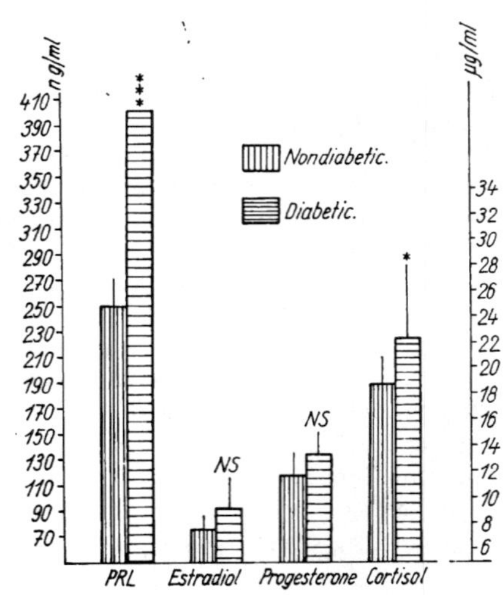

Fig. 2. Serum PRL and progesterone (ng/ml), estradiol (µg/ml) cortisol (µg/dl) in non-diabetic and diabetic pregnant women during the 32nd week

change was manifested during the 32nd as well as the 36th weeks period of gestation. Two hours post glucose oral administration did not cause any further rise in HPL levels.

The exact role of HPL in regulation of glucose tolerance has not yet been fully clarified. It may be acting as a modulator of the maternal intermediary metabolism [9]. A correlation between serum HPL and glucose in blood has been suggested by the observation of *Gaspared* et al. [10] that administration of glucose was followed by a decrease in serum HPL level. Another support for such a relation came from the work of *Tyson* et al. [11] who found a significant rise in HPL level associating hypoglycemia induced by fasting or by insulin.

This suggests that alteration in maternal substrate utilization affected placental function [12]. However other authors obtained controversial results on administration of glucose orally [13].

Human placental lactogen may alter carbohydrate metabolism by diminishgin the effectiveness of maternal insulin and causing marked increased mobilization of FFA. Hence, it was believed to be lipolytic, and diabetogenic [1, 14]. The simultaneous reduction of glucose tolerance and constant rise of HPL throughout pregnancy supports this view. Short term (for 2 h) infusion of HPL did not effect glucose tolerance nor insulin or FFA, but longer term infusion (for 12 h) was followed by impaired glucose tolerance [15]. This suggests that persistant rise of HPL during pregnancy influences glucose tolerance, which may confirm the diabetogenic role of HPL.

In our study of diabetic and non-diabetic pregnant women, E_2 and progesterone levels were not significantly different during the thirty-second week of gestation in both groups. Only E_2 level was significantly increased during the 36th week in the diabetic group. Fasting glucose in the same group was reduced if compared to 32nd week level. Ameliorative effect of natural estrogens and progesterone on diabetes has been reported [16] and might be due to increased hepatic glycogen deposition [17] and/or enhanced insulin secretion [18]. It was, therefore, difficult to classify estradiol or progesterone as insulin antagonists [19]. *Bertoli* et al. [18] suggested that sex hormones might play a role in the control of insulin receptors, and hence might affect glucose homeostasis.

Human growth hormone (HGH) was significantly elevated in diabetics vs non-diabetic contro lgroup. HGHcounteracts insulin binding to its specific receptors [20]. It is known to be diabetogenic due to its lipolytic activity and suppression of glucose utilization.

Plasma cortisol levels rise progressively throughout pregnancy, maintaining the circadian rhythm with little afternoon viariation [21]. Diabetics had higher cortisol levels than the non-diabetic pregnant women ($P < 0{\cdot}05$). In the same cases estradiol was found to be higher in the former group. The increase of cortisol level cannot be attributed for by increase in its secretion rate but may be also due to reduced clearance rate of free and albumin-bound cortisol [22], probably through the effect of hepatic enzymes metabolizing cortisol [23]. Cortisol is known to be a diabetogenic hormone since it counteracts the effectivness of insulin during pregnancy.

Glucose intolerance during GTT was reported inhyperprolactinaemia inspite of exaggerated insulin response [24], and suppression of plasma glucagon [25]. During normal pregnancy, PRL is elevated in serum more than ten times [16]. This suggests a probable role of gestational physiological hyperprolactinaemia in changing glucose tolerance. For this reason we examined both nondiabetic and diabetic pregnant women for PRL. Serum PRL was (M $\pm$ SEM) $252 \pm 23{\cdot}71$ vs $404 \pm 41{\cdot}8$ µg/ml ($P < 0{\cdot}005$) in nondiabetic and diabetic group respectively. HGH, cortisol and glucagon in one side and hyperprolactinaemia in the other side appear to have no synergistic counter-insulin effect [26]. *Posner* et al. [27] reported that HGH and PRL of different species bind to the same receptors in rat liver and called them lactogenic receptor, which were reported to be induced by estradiol and insulin [28]. Binding of PRL to its receptors in liver microsomal membranes from male and female rats was reduced to one third by inducing diabetes [29]. Prolactin was found to activate lipolysis in adipose tissue [30], and therefore it elevates serum FFA which might be the probable the diabetogenic affect of PRL.

The change observed in the levels of different hormones in the 36th, versus the 32nd week of pregnancy were insignificant and were not associated with further deterioration of glucose tolerance. This might indicate that the metabolic machinary is working at its utmost activity during the late weeks of pregnancy, so that no further metabolic changes occurs during this period.

Diabetes in pregnancy may be an exaggeration of the physiological mechanisms that modulate maternal metabolic processes for preservation of substrate requirements for both the fetus and mother.

Received August 17, 1983

References

[1] *W. Uresell, M. Brudenell, T. Chard,* Br. Med. J. **2** (1973) 80

[2] *P. Beck, M. L. Parker, W. H. Daughaday,* J. Clin. Erodocrinol. **25** (1965) 1457

[3] *W. H. Werner, G. Rey, H. Wielinger,* I. Analyt. Chem. **252** (1970) 224

[4] *A. R. Genazzani, F. Cocola, M. Casoli, G. Mello, G. Scarcelli, P. Neri,* J. Obstet. Gynaecol. Br. Commonw. **78** (1971) 577

[5] *G. E. Abraham,* Radioimmunoassay and related procedures in medicine. Int. Atomic Energy Agency, Vienna 1974

[6] *S. M. Glick, J. Roth, R. S. Yalow, S. A. Berson,* Nature **199** (1963) 784

[7] *P. H. Hwange, H. Guyde, H. Friesen,* Proc. Natl. Acad. Sci, **68** (1971) 1902

[8] *R. Garaza, G. E. Abraham,* Analytical letters **5** (1972) 787

[9] *R. Burt, N. H. Leak, A. L. Rhyne,* Obstet. Gynecol **36** (1970) 233

[10] *U. Gaspared, H. Santront, A. Luychx,* J. Obstet. Gynecol. Br. Commonw. **81** (1974) 201

[11] *J. E. Tyson, K. Austin, J. Farinholt, J. Fielder,* Am. J. Obstet. Gynecol. **125** (1976) 1073

[12] *H. Selenkow, K. Varma, D. Younger, P. White, K. Emerson,* Jr., Diabetes **20** (1971) 696

[13] *P. M. Fisher, H. W. Sutherland, P. D. Bewsker,* Diabetologia **19** (1980) 10

[14] *M. D. G. Gillmer, R. W. Beard, N. W. Oakley, F. M. Brooks, M. Brudenell, T. Chard,* Br. J. Obstet. Gynaecol **84** (1977) 197

[15] *P. Beck, W. H. Daughaday,* J. Clin Invest. **46** (1967) 103

[16] *R. K. Kalkhoff, H. J. Kin,* Ciba Found. Symp. **63** (1979) 29

[17] *M. L. Matute, R. K. Kalkhoff,* Erdocrinology **92** (1973) 762

[18] *A. Bertoli, R. Depirro, A. Fusco, A. V. Greco, R. Magnatta, R. Lauro* J. Clin. Endarinol. Metab. **50** (1980) 246

[19] *S. L. Howell, M. Tyhurst, I. C. Green,* Diabetologia **13** (1977) 579

[20] *V. Soman, W. Tamborlane, R. Defrenzo,* N. Engl. J. Med. **299** (1978) 1025

[21] *M. Cohen, M. Stiefel, W. J. Reddy, J. C. Laidlaw,* J. Clin. Endocrinol **78** (1958) 1070

[22] *B. Zumoff, D. K. Kukushima, E. D. Weitzman, J. Kream, L. Hellman,* J. Clin. Endocrinol **39** (1974) 805

[23] *A. Kumagai, M. Otomo, S. Yano, N. Takuchi, K. Bishino, H. Ueda, S. Ko. M. Kitamura,* Endocrinol. Jap. **6** (1959) 85

[24] R. Landgraf, M. M. Langraf-Leur, A. Weissman, R. Horl, P. Scriba, Diabetologia 13 (1977) 99

[25] A. B. Gustafson, M. F. Banasiak, R. Kalkhoff, T. C. Hangen, H. Kim, J. Clin. Endocrinol. Metab. 51 (1980) 242

[26] B. B. Saxena, Endocrinology of Pregnancy. F. Sushs & A. Klopper (eds.) Harper & Row Publishers, Maryland (1977) 222

[27] B. I. Posner, P. A. Kelly, H. G. Friesen. Proc. Natl. Acad. Sci USA. 21 (1974) 2407

[28] P. A. Kelly, B. I. Posner, T. Tsushima, H. G. Friesen, Endocrinology 98 (1974) 532

[29] R. C. Baxter, J. M. Bryson, J. R. Turtle, Metabolism 30 (1981) 211

[30] P. Berle, E. Finsterwalder, M. Apostolakis, Horm. Metab. Res. 6 (1974) 342

Isotopenpraxis 20 (1984) 8, S. 309 bis 311

Effect of Different Doses on Ionizing Radiation on the Distribution of ^{144}Cerium in Rabbits

S. Kossakowski, W. Łojek

(The Veterinary Research Institute, Pulawy, Department of Radiobiology)[1])

Investigations were carried out on 84 rabbits irradiated with doses 103 and 206 mC/kg and next contaminated intravenously with ^{144}Ce 6 days after irradiation. Radioactivity of muscles and internal organs of the rabbits was determined at different periods. It has been found that the course of radiocerium distribution in irradiated rabbits was slower and it remained on a lower level that in the control animals. These changes depended on the dose of ionizing radiation.

Die Untersuchungen wurden an 84 mit Dosen von 103 und 206 mC/kg bestrahlten und anschließend, d. h. 6 Tage nach der Bestrahlung, intravenös mit ^{144}Ce infizierten Kaninchen durchgeführt. Zu verschiedenen Zeitpunkten wurde die Radioaktivität der Muskeln und der inneren Organe der Kaninchen bestimmt. Es wurde festgestellt, daß die Verteilung des radioaktiven Zers bei bestrahlten Kaninchen langsamer verlief und daß die Aktivität auf einem niedrigeren Niveau blieb, als bei den Kontrolltieren. Diese Unterschiede hingen von der Dosis der ionisierenden Strahlung ab.

Keywords

biological localization; biological radiation effects; cerium 144; intestine; rabbits; radiation doses; radionuclide kinetics; X radiation

Introduction

The interest in radiocerium has been prompted by the development of nuclear technology and the possibility of run-away radioactive pollution of the environment. This radioizotope is one of the more hazardous fission/products of ^{235}U, due to its high radioactivity which is manifested by the emission of beta and gamma radiations of energies equal to 0·30 and 0·75 MeV, respectively. Therefore, the study of cerium metabolism in the animal organism is of vital importance especially from the standpoint of the possible utilization of Ce-contaminated animals for human consumption.

Investigations into the fate of radiocerium in the animal organism have shown that 0·05 to 0·1 per cent of its total amount is taken up by the digestive tract [6, 19]. Greater amounts are taken up by the respiratory system [10, 14]. Aqueous and lipid solutions of the isotope penetrate relatively easily through the skin [24]. Radiocerium accumulated mainly in the liver, lungs and bones [17, 23], and may cause postradiation damage [7, 15]. It has also been found to penetrate through the placenta [20] and to be secreted with milk [21].

The above investigations were carried out on healthy animals. However in cases of accidental leaks of radioactive material the exposure to external ionizing radiation is of considerable importance. Diverse and multidirectional effects of ionizing radiation include, among others, metabolite disorders which may affect the metabolism of radioisotopes in an irradiated organism. No data concerning the effect of ionising radiation on the metabolism of radiocerium were found in available literature and it provided the rationale for starting the present investigations.

Experimental

The inevestigations were carried out on 84 rabbits of bot sexes, weighing 3 kg each on average and coming from the same colony. They were fed a standard feed twice a day according to accepted norms. The rabbits wre divided randomly into the following groups:

1. non-irradiated and contaminated with ^{144}Ce (control) — 30 animals
2. irradiated with a dose of 103 mC/kg and contaminated with ^{144}Ce — 30 animals
3. irradiated with a dose of 206 mC/kg and contaminated with ^{144}Ce — 24 animals

The rabbits were irradiated using a Siemens-Bombe X rays apparatus. The following parameters were applied: 180 kV, 20 mA, filter Cu 0·5 mm, dose 0·98 mC/kg/min. A vehicle-free solution of ^{144}cerium chloride made by Nuclear Research Institute, Swierk, was used for contamination. The isotope was administered intraveonusly six days after the irradiation. A single dose of 0·5 cm³ and 1204 Bq radioactivity was applied. All irradiated rabbits were screened for clinical symptoms. The number of leucocytes was counted in irradiated rabbits picked at random on the 6th day after treatment. The following tissues and organs were tested for radiocerium levels: the biceps brachial muscle, quadriceps femoral muscle, blood, brain, lungs, liver, kidneys. Radiometric measurements were made after 1, 3, 7, 15 and 30 days. Six rabbits were examined on each date. The collected samples were incinerated at a tempe-

[1]) Adress: Partyzantów 57, 24-100, Pulawy, Poland

Tab. 1. Radioactivity in Bq of tissues and organs of irradiated rabbits and contaminated with ^{144}Ce

Time of invest.	Doses of irrad.	Thigh muscles	Shoulder muscles	Liver	Kidneys	Heart	Lungs	Brain	Stomach	Large intestine	Blood
1 d	Control	0·80	1·27	173·67	44·21	8·17	63·40	0·53	0·58	0·45	1·33
	103 mC/kg	0·88	1·15	123·11	29·98	6·27	135·25	0·38	5·19	3·87	0·15
	206 mC/kg	0·18	0·27	45·53	8·69	1·00	166·85	0·15	0·92	1·27	0·12
3 d	Control	0·53	0·80	201·87	29·53	5·76	18·51	0·47	1·75	1·42	0·75
	103 mC/kg	0·47	0·80	119·44	17·02	4·74	312·26	0·22	4·87	2·77	0·22
	206 mC/kg	0·18	0·20	38·27	4·65	0·77	211·14	0·15	0·60	0·63	0·12
7 d	Control	0·97	0·97	159·53	24·82	2·80	36·82	0·67	2·67	2·00	0·88
	103 mC/kg	0·70	0·97	184·07	20·52	3·94	353·79	0·45	5·14	3·17	0·38
	206 mC/kg	0·20	0·25	58·85	7·72	1·03	366·65	0·15	1·00	1·33	0·05
15 d	Control	0·53	1·05	168·00	15·22	3·64	8·27	0·38	7·04	5·27	0·33
	103 mC/kg	0·77	0·95	173·41	12·38	4·64	152·42	0·37	3·10	2·30	0·52
	206 mC/kg	0·17	0·22	23·87	22·02	0·77	161·43	0·25	0·60	0·35	0·08
30 d	Control	0·57	0·90	226·41	13·66	3·42	4·94	0·47	5·97	3·15	0·33
	103 mC/kg	0·20	0·33	71·17	5·49	1·70	59·48	0·12	2·75	0·68	0·15

rature of 693 K. Samples of 100 mg each (50 mg from heart) were weighed using an analytical balance with exaanctitude of $\pm 0{\cdot}0001$ g. Radiometric measurements of the samples were made in a lead chamber using a G-M counter type BOH-45 connected with a ZM-701M radiometric set.

Results

The following changes were observed in irradiated rabbits: apathy, reduced mobility, lack of appetite, diarrhoea. The intensity of diarrhoea varied with the irradiation dose. The number of leucocytes in irradiated rabbits was reduced after 6 days by 41·9 and 58·1 per cent following the 103 and 206 mC/kg treatments, respectively.

The results of radiometric measurements of the organ and tissue samples are given in Tab. 1. In non-irradiated rabbits the highest amounts of radiocerium were found after 24 hours in the liver (173·67 Bq), lungs (63·40 Bq) and kidneys (44·21 Bq). On the same date, the lowest amounts were found in the brain, stomach, and large intestine (0·45 – 0·58 Bq). After 3 days, radioactivity increased in the liver (201·87 Bq), stomach (1·75 Bq), and large intestine (1·42 Bq) and decreased in the kidneys (29·53 Bq), and lungs (8·51 Bq). After 7 days the liver, lungs, and kidneys showed the highest radioactivities (159·53, 36·82, and 24·82 Bq respectively), and the blood brain, and muscles showed the lowest ones (0·58, 0·67, and 0·97 Bq respectively). After 15 days, a reduction in radioactivity occured in all samples. The radioactivity was 168·0 Bq in the liver, 15·22 Bq in the kidneys, 8·27 Bq in the lungs, 0·33 to 0·53 Bq in the blood and brain, and 0·53 to 1·05 Bq in the muscles. Finally after 30 days, radioactivity continued to decrease in the lungs (4·94 Bq) and kidneys (13·66 Bq) and sharply in creased in the liver (22·41 Bq). In the remaining samples radioactivity was maintained at the same level as after 15 days.

In rabbits irradiated with 103 mC/kg the measurements made 24 hours after contamination showed the highest radioactivity to occur in the lungs (153·25 Bq), liver (123·11 Bq) and kidneys (29·98 Bq). The lowest radioactivities were found in the brain (0·38 Bq), blood (0·75 Bq) and muscles (0·88 to 1·15 Bq). After 3 days, a marked increase of radioactivity was found in the lungs (312·26 Bq) and a decrease was found in the liver (119·44 Bq) and kidneys (17·02 Bq). After 7 days, a further increase of radioactivity was found in the lungs (353·79 Bq). An increase was also found in the liver (184·07 Bq) and kidneys (29·52 Bq). Radioactivities in the blood and brain ranged between 0·38 – 0·48 Bq, and muscles radioactivity was from 0·70 to 0·79 Bq. After 15 days, a slight decrease in radioactivity was compared to the 7th day values occured in the liver (173·41 Bq) and kidneys (12·38 Bq). A sharp decrease was found in the lungs (152·42 Bq). After 30 day, radioactivity continued to decrease in all samples. The values were 71·17 Bq for the liver, 59·48 Bq for the lungs, 5·49 Bq for the kidneys, 0·15 Bq for the blood, 0·12 Bq for the brain, 0·20 to 0·33 for the muscles.

In rabbits irradiated with 206 mC/kg the highest radioactivity was found after 24 hours in the lungs (166·85 Bq) to be followed by the liver (45·53 Bq) and kidneys (8·69 Bq). Considerably smaller amounts were accumulated in the remaining organs (0·15 to 1·25 Bq). After 3 days, an increase in radioactivity occured in the lungs (211·14 Bq). A reduction was found in the liver (38·27 Bq) and kidneys (4·65 Bq). In the remaining organs radioactivities generally continued at an unchanged level. After 7 days, radioactivity in the lungs continued to rise (366·65 Bq). A rise in radioactivity was also found in the liver (58·86 Bq) and kiedneys (7·72 Bq). After 15 days, radioactivity was found to decrease in the lungs (161·43 Bq) and liver (23·87 Bq) and to increase in the kidneys (22·02 Bq). In the remaining organs radioactivity levels ranged between 0·08 and 0·77 Bq.

Discussion

The distribution of radiocerium in the tissues of non-irradiated rabbits is characterised by hight accumulation in the liver, lungs, and kidneys and low accumulation in the brain, blood and skeletal muscles. It is in agreement with the notion [1] that elements rare earths, cerium being one of them, form unstable compounds with organic acids (citric, lactic, glutamic) and are in this form distributed throughout the organism. The hypothesis may account for the very low cerium concentrations in the muscles. In this connection, it is noteworthy that the concentration of ^{144}Ce in the heart is many times higher than that in the skeletal muscles.

Rare earths elements are known [25,30] to form metalloproteid in the blood. The metalloporteids are rapidly intercepted by Kupfer's cells of the liver and other elements of the reticuloendothelial system (RES) to be stored primarily in organs rich in those elements [28]. It is borne out by experiments which show the rate of the accumulation of ^{144}Ce in the liver to be reduced by the suppression of RES with coloid compounds of carbon [27, 29]. The excretion of cerium by the organism is mainly through the digestive tract and hence its relatively low accumulation in the kidneys [1, 26].

The rabbits were exposed to an irradiation dose of 103 mC per kg considered to be sublethal [2, 5] and to a dose of 206 mC/kg considered to be equal to LD_{50} [11, 22].

The clinical symptoms shown by the rabbits were typical of the subacute and acute postradiation disease [8, 9]. The reduction in the number of leucocytes was in agreement with data by other authors [5, 16].

Characteristic alterations in radiocerium distribution occured in irradiated rabbits. They were manifested primarily by the increased level of the isotope. The increase was correlated with the irradiation dose and attained its peak in the studied organs within 7 days from the application date. At the same time, radiocerium levels in the liver, kidneys and the remaining organs were lower than those in the controls. The exception were rabbits exposed to 103 mC/kg which showed the stomach and intestine levels of ^{144}Ce higher than the controls.

The changes mentioned above are undoubtedly the result of the action of ionizing radiation as well as radiation-related functional disorders and morphological injures. This is primarily true of the gigestive tract which is particularly sensitive to ionizing radiation [4]. Inflammatory lesions developed in the digestive tract and inflammation-related defective take-up and enhanced excretion undoubtedly contributed to the reduced radiocerium levels in the liver and kidneys. The effect of ionizing radiation on RES, which impaire its coloidal activity depending on the irradiation dose and the severity of postradiation damage is also vital significance in this respect [12]. The increased accumulation of radiocerium in the lungs is related to postradiation swellings and inflammations in that organ [3] as well as to postradiation haemodynamic disorders [13, 18].

Received September 14, 1983

References

[1] *A. Aeberhardt, P. Nizza, Y. Arnoud*, Bull. Inf. Sci. Tech. Commis. Energie Atom. **47** (1961) 2
[2] *S. B. Balmuchanov, A. G. Sergazin*, Javlenije pronicajemosti v rozvitii luczevych poraženij. Alma Ata 1968
[3] Commitee on Pathologic Effects of Atomic Radiation. Science **124** (1956) 60
[4] *R. A. Conard*, Radiat. Res. **41** (1956) 167
[5] *Z. Dienstbier, M. Arient*, Ekshperimentalni postiradiacni syndrom. Praha 1966
[6] *R. Garner*, J. Agric. Sci. **55** (1960) 1
[7] *F. Hahn, D. Lundgreen, R. McClellan*, Radiat. Res. **82** (1980) 1
[8] *S. A. Kirillov, J. V. Vissonov*, Med. Radiol. **12** (1968) 20
[9] *V. I. Kirishkin, V. N. Deshchenko, Z. B. Podgorodecka*, Med. Radiol. 8 (1963) 33
[10] *T. V. Korolev*, Med. Radiobiol. **1967**, 7
[11] *S. Kossakowski*, Medycyna Wet. **24** (1968) 449
[12] *S. Kossakowski, W. Lojek*, Pol. Arch. Wet. **16** (1973) 753
[13] *S. Kossakowski, W. Lojek, J. Kuczynski*, Pol. Arch. Wet. **17** (1974) 633
[14] *D. Lundgreen*, Radiat. Res. **58** (1974) 3
[15] *D. Lundgreen, F. Hahn, R. McClellan*, Health Phys. **38** (1980) 4
[16] Manual on Radiation Hematology,IAEA Vienna 1971
[17] *R. McClellan*, Radiat. Res. **62** (1975) 3
[18] *A. Morgan, C. M. Arkell*, Nature **191** (1961) 1100
[19] *J. I. Moskalev*, Sbornik referatov po radiacjonnoj medicinie. Moskva (1961) 4
[20] *F. Mraz*, Radiat. Res. **64** (1975) 2
[21] *A. Naharin*, Health Phys. **27** (1974) 2
[22] *V. Nair*, Radiat. Res. **30** (1967) 359
[23] *H. Neshev*, I-aja Radiobiol. Konf. Socjal. Stran, Spindlerov Mlyn 1974
[24] *S. Rogaczeva*, Rozpredelenije biologiczeskoje dejstwije i migracja radioaktivnych izotopow. Moskva 1961
[25] *B. Rosoff, H. Spencer*, Health Phys. **28** (1975) 611
[26] *D. Sklan, D. Dubrov, U. Eisner*, J. Nutr. **105** (1975) 1549
[27] *V. Slouka*, Pharmacology **2** (1969) 357
[28] *I. B. Tokin*, Arch. Anat. Gist. Embriol. **58** (1970) 45
[29] *T. Trnovec, A. Pleskova, D. Chorvat, M. Vladar*, Strahlentherapie **143** (1972) 570
[30] *N. A. Zapolskaja, A. V. Fedorova, T. M. Vojevodina*, Rept. Staatl. Zentr. Strahlenschutz **157** (1974) 84

Isotopenpraxis 20 (1984) 8, S. 312 bis 314

Effect of Gamma Radiation on the Electrical Resistivity of Some Metal Acetylacetonate Complexes

N. Kadry, A. Sawaby, A. El Agrami, R. Sharkawi

(Faculty of Science, Menia University, El-Menia, Egypt)

The effect of gamma radiation on the electrical resistivity of some lanthanide and alkali acetylacetonate complexes was studied for differet doses (20·49 – 36·2 Mr). The activation energy for each complexes was calculated. The obtained results showed that the resistivity as well as the activation energy increase with increasing the radiation dose. An empirical linear relation between the rate of change of resistivity with radiation dose and the ionic radius of the metal was obtained.

Der Einfluß von Gammastrahlung auf den elektrischen Widerstand einiger Lanthanid- und Alkali-Azetylazetonat-Komplexe wurde unter Anwendung unterschiedlicher Dosen (20,49 – 36,2 Mr) untersucht. Für jeden Komplex wurde die Aktivierungsenergie errechnet. Die Ergebnisse zeigten, daß der Widerstand und die Aktivierungsenergie mit steigender Bestrahlungsdosis anwachsen. Es wurde eine empirische Beziehung zwischen der Widerstands-Änderungsrate und der Strahlungsdosis sowie dem Ionenradius des Metalls ermittelt.

Keywords

activation energy; alkali metal complexes; electric conductivity; gamma radiation; lanthanum complexes; radiation doses

Introduction

Acetylacetenate is a well known chelating agent [1]. Some authors reported about the effect of ionizing radiation on metal chelated compounds [2 – 4]. *Aly* et al. [2] studied the effect of γ-radiation on the degradation of the M-O bond of some trivalent acetylacetonate complexes. The effect of radiation on the IR spectra of lanthanide 8-hydroxyquinoline complexes was previously studied [5]. Moreover, the induced electrical conductivity of gamma irradiated 8-hydroxyquinoline metal complexes was reported [4].

In the present work, some lanthanide (Ce, Sm, Eu, Gd, Ho, Er) and alkali metal (Na and K) acetylacetonate complexes were investigated before and after exposing to gamma radiation of different doses at different temperature. Study of the dependence of the induced electrical resistivity of the irradiated complexes on the type of the metal bond was also dealt with.

Experimental

The metal acetylacetonate complexes were prepared as reported previously [1]. The elemental chemical analysis was carried out by the Microanalytical Unit, National Research Centre, Cairo. A general chemical formula of $M(Ac,Ac)_3$ was found for the lanthanide investigated complexes.

The electrical resistivity of the investigated complexes were measured on discs of 13 mm diameter and 2 mm thickness. The discs were prepared under a pressure of 200 K bar/cm^2. The cell used in this work was reported before [6] and the current was measured using a 600 B Keithly type electrometer. Measurement of temperature was carried out using a copper-constantan thermocouple placed close to the sample.

The samples were irradiated by //Co gamma 3500-NORATON. The position of irradiation was calibrated by ferrous sulphate dosimeter and was found to give a dose rate of $1{\cdot}9 \times 10^5$ R/h $\pm$ 5% in February 1981.

Results and Discussion

Tab. 1 shows the calculated electrical resistivity of the investigated compounds before and after γ-irradiation with doses 20·44, 22·62 and 36·8 Mr. From this Tab. 1, it is clear that at room temperature in general, the values of

Tab. 1. Specific electric resistivity of alkali andlanthanide metal acety lacetonate complexes at different doses

Element	The resistivity at different doses $\times 10^{-10}$ Ωcm			
	0	20·44 Mr	28·62 Mr	36·8 Mr
Ce	1·33	1·46	1·58	1·70
Sm	1·52	1·64	1·73	1·83
Eu	1·75	1·80	1·92	2·11
Gd	3·04	3·15	3·42	3·84
Ho	1·92	2·02	2·29	2·46
Er	2·21	2·34	2·40	2·53
Na	0·65	0·75	0·84	0·96
K	0·55	0·65	0·76	0·83

Tab. 2. The activation energy for alkali and lanthanide metal acetylacetonate compleces at different doses

Element	The activation energy (e.v.) at different doses			
	Zero dose	20·44 Mr	28·62 Mr	36·8 Mr
Ce	2·978	3·063	3·129	3·176
Sm	3·088	3·104	3·205	3·375
Eu	3·176	3·182	3·294	3·414
Gd	3·573	3·643	3·590	3·812
Ho	3·335	3·568	3·416	3·494
Er	3·465	3·591	3·602	3·706
Na	1·815	1·826	1·877	1·906
K	(0·794	(0·809	(0·835	(0·874
	(0·425	(0·441	(0·454	(0·485

Table 3. The rate of change of resistivity with radiation dose.

Element	Ce	Sm	Gu	Gd	Ho	Er	Na	K
$L \times 10^3$ Mr^{-1}	6·216	4·973	3·73	4·351	6·84	3·73	9·32	11·2

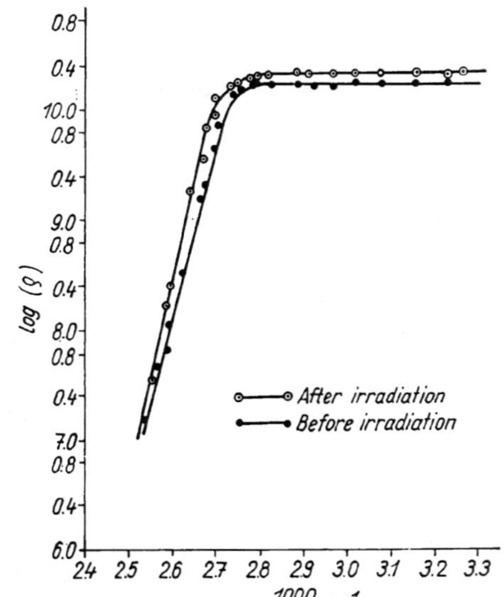

Fig. 1. Relation between log(ϱ) and $\frac{1000}{T}$ K^{-1} for Eu(acac)$_3$ complex

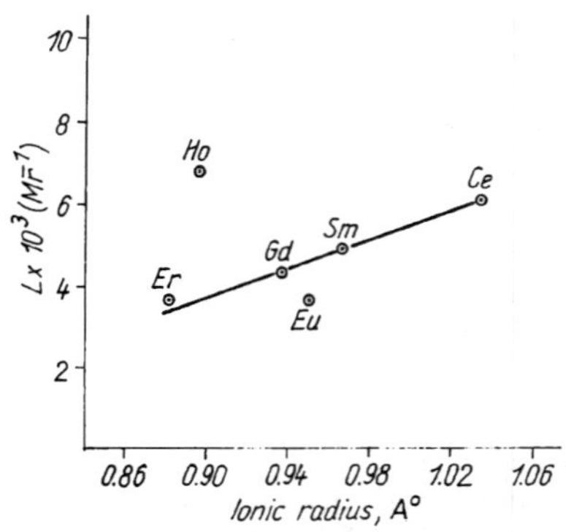

Fig. 3. Relation between ($L \times 10^3$ Mr^{-1}) and the ionic radius fo lanthanide metal cation

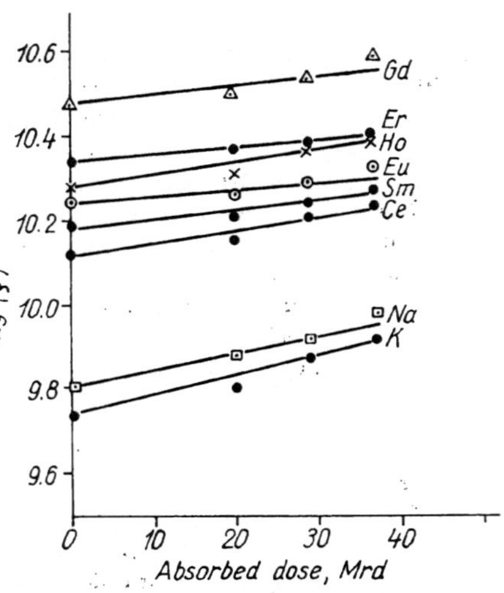

Fig. 2. Relation between log(ϱ) and the absorbed dose (M.r.)

resistivity for the investigated metal complexes increases with the increase of radiation dose. This behaviour can be explained according to *Vinyard* et al. [7]. They proposed that the primary effect of ionizing radiation leads to the production of positive metal ions and electrons. Under the influence of the field produced by the surrounding lattice ions, it moves out into an interstitial position. In this way, a considerable number of negative vacancies may be greated which introduce impurity levels. These levels serve as recombination levels for electrons and thus case a considerable decrease in current and subsequent increase in the resistivity.

The electrical resistivity of the irradiated samples were measured at different temperatures ranging from room temperature up to 120 °C. Fig. 1 shows an example of the obtained relation between log(φ) and 100 T^{-1} for Eu (acac)$_3$ complex. The activation energies were calculated from such figures and then tabulated in Tab. 2. From this table, it is clear to show that the activation energy increase with increasing the radiation dose. This indicates that the mobile π electrons hypothesis [8] remains valid as long as it concerns separate molecules.

The rate of change of resistivity (L Mr^{-1}) with radiation dose is calculated from the following equation:

$$2{\cdot}303 \log(\varrho_D/\varrho) = L.\ D$$

where, L in Mr^{-1}, ϱ and ϱ_D are the resistivities before and after gamma radiation dose D. The values of L were obtained from the slopes of the lines showed in Fig. 2, and then tabulated in Tab. 3.

Fig. 3 shows the resulted linear relation between $L(Mr^{-1})$ and the ionic radius of lanthanide metal cation which means that the degradation process depends on the ionic radius of the metal. This behaviour indicates that the increase in the resistivity with the dose is mainly due to the tupture of the metal ligand bonds.

Received September 20, 1983

References

[1] *F. J. Welcher*, Organic Analytical Reagents. Vol. 1, 1955
[2] *H. F. Ali, F. M. Abdel Kerim, M. M. El-Dessouky, A. M. El-Agrami*, Radichem. Radioanal. Letters 28 (1977) 109
[3] *H. F. Ali, F. M. Abdel Kerim, A. Wl-Agrami, A. M. Atya*, Isotopenpraxis 13 (1977) 59
[4] *A. M. El-Agrami, H. F. Ali, F. M. Abdel Kerim, M. H. Atya*, Isotopenpraxis 15 (1979) 223
[5] *A. M. El-Agrami, F. M. Abdel Kerim, H. A. Ali, M. H. Atya*, Isotopenpraxis 15 (1979) 108
[6] *H. F. Ali, F. M. Abdel Kerim, H. H. Afifi*, Z. Naturforsch. 319 (1976) 675
[7] *J. Vinyard, J. Piennes*, Radiation effect in solids, Moscow 1960
[8] *S. J. Ashcroft, G. Beech*, Inorg. Thermodynamic, Van Norstrand Reinhold Company Ltd, N.Y., 1973

Isotopenpraxis 20 (1984) 8, S. 314 bis 315

Kurze Originalmitteilungen

Untersuchungen zur ^{15}N-Markierung der in einem zellfreien Translationssystem *de novo* synthetisierten Proteine

A. Schröder, B. Löhrke

(Akademie der Landwirtschaftswissenschaften der DDR, Forschungszentrum für Tierproduktion Dummersdorf-Rostock)[1])

Keywords

cell-free translation; labelled compounds; leucine; nitrogen 15; proteins; tritium compounds

1. Einleitung

Zellfreie Translationssysteme werden zur Lösung molekularbiologischer und ernährungsphysiologischer Fragestellungen bezüglich der Proteinbiosynthese herangezogen.

Qualitative und quantitative Untersuchungen mit Hilfe zellfreier Translationssysteme wurden bisher ausschließlich mit radioaktiv markierten Tracern (^3H, ^{14}C, ^{35}S-markierten Aminosäuren) [1–4] durchgeführt, weil deren Nachweis bzw. Messung auch in größeren Meßserien relativ einfach flüssigkeitsszintillationstechnisch vorgenommen werden kann. Nachteilig ist, daß radioaktive Präkursormoleküle oft instabiler sind und daß besondere Sicherheitsanforderungen bei ihrer Handhabung und bei der Abfallbeseitigung beachtet werden müssen. Deshalb wurde geprüft, ob für die im Mikrolitermaßstab arbeitenden Translationssysteme eine Markierung der *de novo* synthetisierten Proteine mit ^{15}N-markierten Aminosäuren als Präkursoren quantitativ auswertbare Resultate ergibt.

2. Methoden

Die Präparation der Zellsaftfraktion aus Rattenleber wurde nach *Junghahn* [1] durchgeführt. Die Polysomen wurden nach *Löhrke* [5] gewonnen.

Das zellfreie Translationssystem enthält in 100 µl Endvolumen (Endkonzentrationen) 175 mM RNase freie Saccharose; 50 mM Tris. Acetat pH 7,6; 4 mM Magnesiumacetat; 40 mM Kaliumacetat; 1 mM ATP; 0,2 mM GTP; 3 mM Phosphoenolpyruvat; 0,8 µg Pyruvatkinase; im Falle der radioaktiven Markierung 2,4 nMol der 19 proteinogenen L-Aminosäuren außer L-Leucin, von dem 37×10^3 Bq als ^3H-Leucin (s. A. 1768 GBq) eingesetzt wurde bzw. im Falle der ^{15}N-Markierung 2,4 nMol der proteinogenen ^{15}N-Aminosäuren einschließlich ^{15}N-Leucin; 15–20 µg Polysomen; sowie 20, 40 und 80 µg Protein der Leberzellsaftfraktion. Inkubiert wurde 40 und 80 min bei 37 °C. Als Kontrolle diente der Ersatz der Leberzellfaftfraktion durch Aqua dest. Die Reaktion wurde durch Zugabe von 100 µl eiskalter 10%iger Trichloressigsäure mit 1 mM DL-Leucin abgestoppt, die Proben 15 min bei 90 °C hydrolysiert, das Präzipitat auf Glasfaserfilter (Whatmann GF/C) gesammelt, getrocknet und in einer Toluol-Szintillatorlösung (1 l Toluol mit 4 g PPO, 0,1 g POPOP und 10 mM Hyaminhydroxid) ausgezählt (LKB Wallace Liquidszintillationscounter).

Zur Messung der ^{15}N-Anreicherung wurden die Filter nach Trocknung unter Vakuum ($<1,33 \times 10^{-1}$ Pa) und Zusatz eines Oxydationsgemisches (CuO/CaO 1:1 m%) in ^{15}N-Entladungsröhrchen übergeführt, um die Proteine direkt zu oxydieren [6,7]. Anschließend erfolgte die Auswertung mit dem ^{15}N-Analysator NOI-5 (VEB STATRON). Der N-Gehalt wurde über die im zellfreien Translationssystem eingesetzten Stammlösungen und den Proteingehalt der Zellsaftfraktion ermittelt. Die Proteinbestimmung erfolgte nach *Bradford* [8]. Der Nucleinsäuregehalt der Polysomen wurde in die Bilanz mit einbezogen, ausgehend davon, daß 1 µg Polysomen einer Adsorption von 13,4 Einheiten bei 260 nm entsprechen. Zur Auswertung gelangten je 3 Untersuchungen, wobei der Fehler des Mittelwertes in den Abbildungen mit ausgewiesen ist.

3. Ergebnisse und Diskussion

Ein Vergleich der Einbauraten von ^3H-Leucin und ^{15}N-Aminosäuren in Peptide ergibt, daß sehr ähnliche Kurvenverläufe erhalten werden (Abb. 1, 2). Sowohl bei ^3H-Leucin- als auch bei ^{15}N-Markierung nimmt die Einbaurate mit zunehmender Inkubationszeit zu (Abb. 1, 3b). An-

[1]) Anschrift: DDR-2551 Dummerstorf

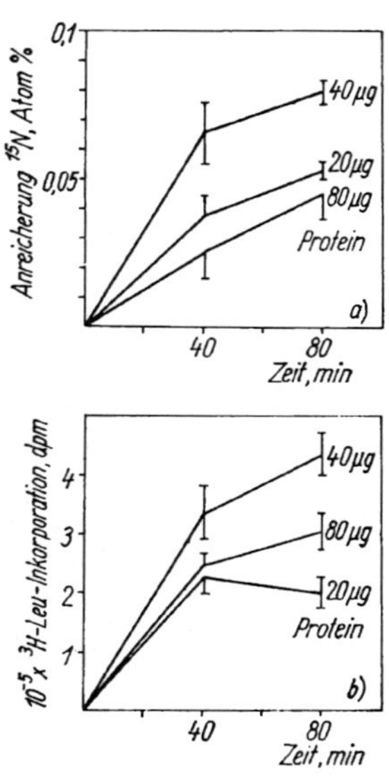

Abb. 1

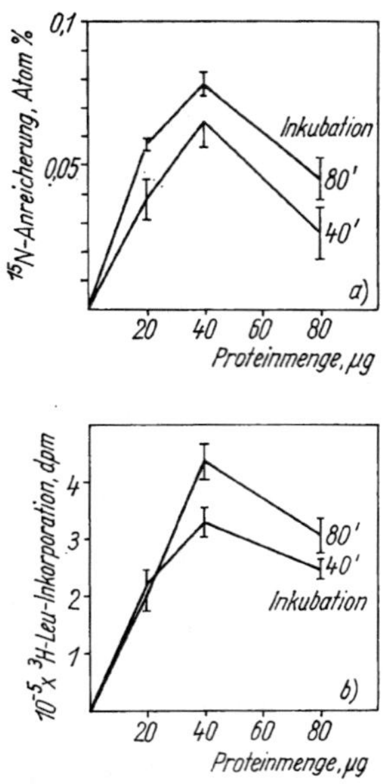

Abb. 2

Abb. 1. Abhängigkeit der Aminosäureinkorporation von der Inkubationszeit
a) ^{15}N-Markierung; b) ^{3}H-Leucin-Markierung

Abb. 2. Abhängigkeit der Aminosäureinkorporation von der Proteinmenge der Zellsaftfraktion
a) ^{15}N-Markierung; b) ^{3}H-Leucin-Markierung

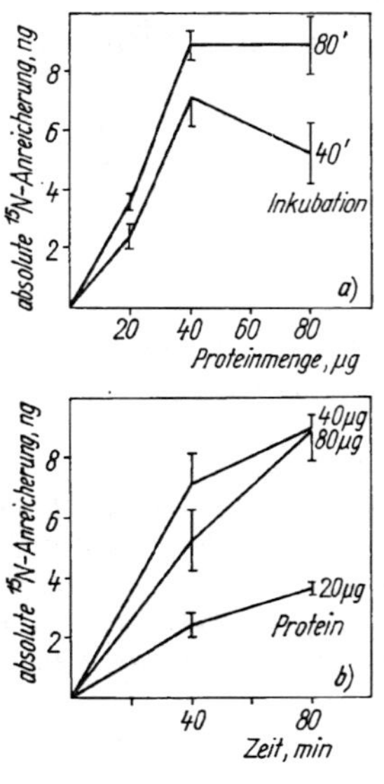

Abb. 3. Darstellung des absoluten ^{15}N-Einbaus in Abhängigkeit von der Proteinmenge der Zellsaftfraktion (a) und von der Inkubationszeit (b)

dererseits differieren die Kurvenverläufe beider Markierungsverfahren, wenn die konzentrationsabhängige Translationsrate (Abb. 2) betrachtet wird. Die Abhängigkeit des Aminosäureeinbaus von der Proteinkonzentration der Leberzellsaftfraktion ist in den Abb. 2 und 3a dargestellt. Bis etwa 40 µg eingesetzten Zellsaftproteins ist ein näherungsweise linearer Anstieg der Kurven zu erkennen. Bei höherer Proteinkonzentration der Zellsaftfraktion (80 µg) ist der Aminosäureeinbau wieder reduziert. Unter Berücksichtigung der Isotopenverdünnung durch die dem Translationssystem zugesetzte Zellsaftproteinmenge kann erklärt werden, daß trotz der hohen Einbaurate von ^{3}H-Leucin (Abb. 1) bei Zusatz von 80 µg Zellsaftprotein die geringsten Anreicherungen des Isotops ^{15}N gemessen werden. Erst ein Bezug der angegebenen Anreicherungen des Isotops ^{15}N auf die im System befindliche N-Gesamtmenge erlaubt den direkten Vergleich der Befunde, die durch radioaktive und stabilisotope Markierung gefunden wurden. Deutlich zum Ausdruck kommt der durch hohe Proteinkonzentration verursachte Isotopenverdünnungseffekt (Abb. 2a, 3a) bei der ^{15}N-Analyse. Für die in Abb. 1a dargestellte Kurve mit 80 µg Zellsaftprotein wurden die geringsten ^{15}N-Anreicherungen gemessen, während für diese Konzentration bei radioaktiver Markierung höhere Einbauraten im Vergleich zu 20 µg Zellsaftprotein erhalten wurden (Abb. 1b). Demnach geben die gemessenen relativen Häufigkeiten des Isotops ^{15}N keine erschöpfende Auskunft über die Kinetik des verwendeten polysomalen Translationssystems.

Um den direkten Vergleich zu den Aussagen der ^{3}H-Leucinmethode zu ermöglichen, wurde die absolut in die durch die Translation gebildeten Polypeptidketten eingebaute Menge des Isotops ^{15}N berechnet (Abb. 3).

Um ^{15}N-Anreicherungen nachweisen zu können, müssen mindestens 0,015 Atom % Anreicherung vorliegen, damit die Fehlergrenze der Methode überschritten wird.

Wird die Anreicherung auf die Gesamt-N-Menge bezogen, bedeutet das für die ^{3}H-Leucininkorporation, daß mindestens 80000 dpm (für 40 µg Zellsaftprotein) erreicht werden müssen. Das bedeutet, daß bei der angegebenen Konzentration Zellsaftprotein 0,1 Atom % ^{15}N angereichert werden können, was etwa 9 ng des Isotops ^{15}N in durch die Translation neu gebildeten Peptiden entspricht. Bei Einsatz einer geringeren Menge Zellsaftprotein können die gemessenen Anreicherungen in Atom % höher sein, entsprechen aber einer geringeren Menge des inkorporierten Isotops ^{15}N. Die aus den ^{15}N-Anreicherungen errechneten Mengen inkorporierter Aminosäuren sind höher als die mit ^{3}H-Leucin erhaltenen Ergebnisse ausweisen, was damit zu erklären ist, daß ^{3}H-Leucin in nur etwa 100fachem Unterschuß im Verhältnis zu den übrigen proteinogenen Aminosäuren dem Translationssystem zugesetzt wurde.

Aus Kostengründen wurde ein äquimolarer Einsatz der radioaktiv markierten Aminosäuren nicht erwogen.

Eingegangen am 5. 10. 1983

Literatur

[1] *I. Junghahn, J. Scholz,* Acta. biol. med. germ. **31** (1973) 691
[2] *P. A. Rogers, G. H. Jones,* Can. J. Biochem. **60** (1982) 587
[3] *W. Zagorski,* Anal. Biochem. **87** (1978) 316
[4] *K. Marcu, B. Dudock,* Nucl. Acids Res. **1** (1974) 1385
[5] *B. Löhrke,* Wachstumsleistungsfrüherkennung mit Hilfe zellfreier Translationssysteme. Forschungsbericht, Akademie der Landwirtschaftswissenschaften der DDR, Forschungszentrum für Tierproduktion Dummerstorf-Rostock, 1981
[6] *H. Faust,* Isotopenpraxis **3** (1967) 100
[7] *K. Wetzel,* Isotopenpraxis **10** (1974) 85
[8] *M. Bradford,* Anal. Biochem. **72** (1976) 248

Tagungsberichte

Symposium „Aktuelle Ergebnisse der ^{15}N-Tracertechnik in der Medizin" vom 29. Mai bis 1. Juni 1983 in Rostock

Vom 29. Mai bis 1. Juni 1983 trafen sich auf Einladung der Kinderklinik der Wilhelm-Pieck-Universität Rostock (Prof. *W. Heine*) und der Arbeitsgemeinschaft Stabile Isotope des Zentralinstituts für Isotopen- und Strahlenforschung der AdW der DDR Leipzig (Prof. *H. Faust*) 70 Wissenschaftler aus 20 Einrichtungen der DDR, der BRD, der UdSSR, der Schweiz und Kanadas, um auf der Grundlage von 32 Vorträgen „Aktuelle Ergebnisse der ^{15}N-Tracertechnik in der Medizin" zu diskutieren. Das Programm umfaßte die vier Themengruppen

- Methodisch-analytische Aspekte der ^{15}N-Tracertechnik,
- Mathematische Behandlung und Interpretation der Versuchsdaten,
- ^{15}N in der Gastroenterologie und Ernährungsforschung,
- ^{15}N in der Endokrinologie und in anderen Fachgebieten.

In einer einleitenden Bestandsaufnahme stellte *W. Heine* (Rostock) fest, daß die ^{15}N-Tracertechnik heute potentiell ein bedeutsames Werkzeug der experimentellen Forschung in der Medizin darstellt und als eine den Patienten nicht belastende Technik enorme Möglichkeiten bietet. Zu den führenden Zentren in der Welt sind Baltimore, Cambridge/Mass., Leipzig, London, Toronto und Zürich zu zählen. In der DDR besteht daneben eine aktive Arbeitsgruppe in Rostock.

Zur Methodik der ^{45}N-Tracertechnik in der Medizin stellte *H. Faust* (Leipzig) zusammenfassend einen anspruchsvollen komplexen Tracerversuch vor, bei dem ein gesunder Proband ein Jahr lang bezüglich seines Ganzkörper-Stickstoffstoffwechsels nach ^{15}N-Glycin-Markierung unter zahlreichen Gesichtspunkten beobachtet wurde. Teilaspekte solcher Untersuchungen trugen die Leipziger *H. Bornhak* (Plasmaprotein-Turnover), *K. Hirschberg* (Analytik und Turnover der Nicht-Protein-Stickstoffverbindungen), *K. Jung* (^{15}N-Isotopenanalytik) und *P. Junghans* (mathematische Interpretation tracerkinetischer Daten) vor. Zu methodischen Aspekten sprachen außerdem *K. Krawielitzki* (Rostock-Dummerstorf), *K. Wutzke* (Rostock) und *J. Töwe* (Rostock).

P. B. Pencharz (Toronto) gab einen interessanten Überblick über seine ^{15}N-Arbeiten der letzten 9 Jahre zur Ernährungsforschung an Kindern und ging dann ausführlich auf neue Ergebnisse zu spezifischen Ernährungsproblemen ein. Vergleichende Untersuchungen zur Ernährung Frühgeborener stellte *Chr. Plath* (Rostock) vor. *J. Putziger* (Leipzig) untersuchte die Verwertung verschiedener Aminosäuren. Zur Ernährungsforschung sind auch die ^{15}N-tiermedizinischen Arbeiten von *G. Gebhardt* (Leipzig), *H. Bergner* (Berlin), *O. Simon* (Berlin) und *R. Münchmeyer* (Berlin) zu zählen, die beispielhaft Stickstoffbedarf und -verwertung behandelten.

Zu gastroenterologischen ^{15}N-Untersuchungen des Proteinstoffwechsels sprachen *W. Hartig*/Leipzig (Proteinumsatz im Postaggressionsstoffwechsel), *H.-D. Czarnetzki*/Rostock (Proteinmetabolismus nach Magenoperationen), *H. J. Stolpe*/Rostock (Intestinaler Harnstoffstoffwechsel), *F. Walter*/Leipzig (Proteinsynthese bei Gastrektomierten), *R. Matkowitz*/Leipzig (Proteinstoffwechsel im Coma hepaticum) und *F. Gehrandt*/Pritzwalk (Proteinstoffwechsel vor und nach Leber-Shuntoperationen).

P. Krumbiegel (Leipzig) stellte eine Methode zur Charakterisierung der Entgiftungsfunktion der Leber mit Hilfe ^{13}C- oder ^{15}N-markierter Pharmaka zur Diskussion.

Zum Themenkomplex „^{15}N in der Endokrinologie" referierte *M. Zachmann* (Zürich) über den Einfluß von Östrogenen auf die ^{15}N-Bilanz im Harn von Mädchen. Er fand einen protein-anabolen Effekt, der aber geringer ist als der der früher untersuchten Hormone HGH und Testosteron. *H. Willgerodt* (Leipzig) sprach zum Einfluß der Schilddrüsenfunktion auf die Stickstoffbilanz und zum Einsatz von ^{15}N-Aminosäuren für die Differentialdiagnostik des Pseudohermaphroditismus masculinus. *I. Richter* (Rostock) hielt einen Vortrag zum Stickstoffmetabolismus bei minderwüchsigen Kindern.

Den Einfluß eines Myocardinfarkts auf die Bildung von Glutamin und weiteren Verbindungen untersuchte *O. I. Pisarenko* (Moskau) mit Hilfe von ^{15}N-Ammoniak am isolierten Herzmuskel.

Weitere Vorträge galten ^{15}N-Untersuchungen zum Streßeinfluß auf die Proteinsynthese im Tierversuch (*G. Pietsch*/Leipzig: Schwere Verbrennungen; *Chr. Plath*/Rostock: Pränatale Hypoxie) sowie der Residualvolumenbestimmung der menschlichen Lunge (*R. Reinhardt*/Leipzig) und der Synthese ^{15}N-markierter Aminosäuren (*H. Engelmann*/Berlin).

Alle Vorträge wurden als Band **77** (1983) der ZfI-Mitteilungen veröffentlicht; Freiexemplare können bei Prof. *H. Faust*, Zentralinstitut für Isotopen- und Strahlenforschung der ADW 7050 Leipzig Permoserstr. 15 angefordert werden.

Das zweite ^{15}N-Symposium zu dieser Thematik ist für 1985 in Leipzig und das dritte für 1987 in Ahrenshoop vorgesehen.

P. Krumbiegel

Eingegangen am 7. 2. 1984

Isotopenpraxis 20 (1984) 8, S. 317 bis 320

Buchbesprechungen

Studien zur Entwicklungsgeschichte der Kernphysik in der UdSSR

Autorenkollektiv. Kiew: Naukowa dumka 1982. 322 S., zahlr. Abb. und Tab. (in Russisch); geb. Rbl. 4; M 20,—

Das vorliegende Buch stellt 26 Beiträge aus der Kernphysik und aus angrenzenden Wissenschaften vor. Es erschien anläßlich des 50. Jahrestages der ersten in der UdSSR (Wiederholung des berühmten Experiments von *J. D. Cockcroft* und *E. T. S. Walton*) durchgeführten Spaltung des Lithiumkerns durch beschleunigte Protonen im Physikalisch-Technischen Institut in Charkow.

Der Inhalt des Werkes ist bedeutend breiter angelegt, als man aus dem Titel herauslesen kann. Neben der historischen Exkursion in die Zeit der Entstehung und Entwicklung der Kernphysik in der UdSSR — vorgestellt im wesentlichen durch 6 Beiträge — werden aktuelle Probleme der Kernphysik, der Elementarteilchenphysik, der thermonuklearen Synthese und der Kernenergetik behandelt.

Im Artikel „Kernphysik der 30er Jahre" von *D. D. Ivanenko* wird sozusagen aus erster Hand geschildert; der Autor war selbst unmittelbarer Teilnehmer der kernphysikalischen Ereignisse dieser Jahre und beschreibt die geistige Atmosphäre, in der die Entwicklung der Kernphysik vor sich ging. Für die deutschen Leser ist dabei sicher von Interesse, daß die schon 1934 von *Ida Noddack* ausgesprochene Idee über die mögliche Spaltung schwerer Kerne unter Neutroneneinfluß eine gebührende Anerkennung durch den Autor erfährt.

Der Artikel von *N. A. Perfilow, K. A. Petrschak* und *W. P. Eismont* ist der Entstehung der Kernphysik in der UdSSR gewidmet. Das Radium-Institut in Leningrad besaß schon Mitte der 20er Jahre umfangreiche Detektionsmethoden für die Kernstrahlung. Hier entstand auch die Idee für den Bau eines Zyklotrons, und der damalige Mitarbeiter des Institutes *G. A. Gamow* gab die richtige theoretische Interpretation des α-Zerfalls auf der Grundlage der gerade entstandenen Quantenmechanik.

Im Beitrag von *A. A. Zigikalo* und *A. P. Klutscharew* „Elektrostatische Beschleuniger des Physikalisch-Technischen Institutes in Charkow" sind die Arbeiten zur Gewinnung beschleunigter geladener Teilchen zur Untersuchung der Kerne zusammengefaßt, die dann 1932 zur Spaltung des Lithiumkerns durch beschleunigte Protonen geführt hatten.

Großen Anteil an der Herausbildung der Kernphysik in der UdSSR hatte auch das Leningrader Physikalisch-Technische Institut. *A. P. Grinberg* beschreibt in „I. V. Kurtschatow und die ersten Beschleuniger im Leningrader Physikalisch-Techschnischen Institut" lebendig die Geschichte des Aufbaus des ersten Zyklotrons unter der Leitung von *I. V. Kurtschatow*.

Die Thematik der Artikel, die sich mit dem modernen Stand der Theorie und des Experiments der Kernphysik befassen, gibt einen Einblick in den heutigen Problemkreis dieser Wissenschaft. Quantenelektrodynamik, Kernmodelle, Anregungszustände der Kerne, Diffraktionserscheinungen, die Kerne an den Grenzen ihrer Stabilität, relativistische Kernphysik, Struktur der Hadronen, Kernreaktionen, Beschleunigertechniken, Physik der Kernspaltung, das sind die angesprochenen Themenkreise. Schon aus dieser Aufzählung ist zu ersehen, daß sich das Buch nicht nur an die an der Geschichte der Kernphysik interessierten Leser wendet, sondern auch an diejenigen, die auf diesem Gebiet aktiv tätig sind.

N. Gorski

Guide-Lines to Planning Atomic Spectrometric Analysis Studies in Analytical Chemistry Vol. 4

B. Magyar. Budapest: Akadémiai Kiadó 1982. 273 S., 67 Abb., 39 Tab.; gebunden M 62,50

Dem Charakter dieser Reihe entsprechend werden im vorliegenden Band weniger Details analytischer Prozeduren beschrieben, worüber bereits ausführliche Abhandlungen erschienen sind, sondern es werden vielmehr die physikalisch-chemischen Hintergründe der Atomabsorptions-Spektrometrie (AAS) und Atomemissions-Spektrometrie (AES) im Zusammenhang mit ihren praktischen Konsequenzen diskutiert. Durch ein tieferes Verständnis dieser Methoden soll — wie es im Titel des Buches zum Ausdruck kommt — für den Anwender die Wahl des zur Lösung eines gegebenen analytischen Problems am besten geeigneten Vorgehens erleichtert werden.

Kap. 1 bringt einen Vergleich der Leistungsfähigkeit der wichtigsten physikalischen Methoden der Elementanalyse, eine Retrospektive optischer Methoden sowie eine kurze Erläuterung des Prinzips der Spektrometrie.

In Kap. 2 werden Eigenschaften, Erzeugung und Absorption von elektromagnetischer Strahlung behandelt.

Kap. 3 befaßt sich mit den Ursachen für die Verbreiterung der Spektrallinien, mit den unterschiedlichen Linienformen und den analytischen Konsequenzen der Linienverbreiterung.

Kap. 4 erläutert den Zusammenhang zwischen Atomspektrum und Elektronenstruktur. Es handelt sich praktisch um einen kurzen Abriß der Theorie der Atomspektren und führt letztendlich zu den Auswahlregeln sowie zu den Energieniveau-Diagrammen von Vielelektronenatomen.

Das Kap. 5 enthält zahlreiche neue experimentelle Ergebnisse, die eine Abschätzung der Empfindlichkeit der verschiedenen AAS- bzw. AES-Techniken ermöglichen. Dieses sehr umfangreiche und instruktive Kapitel führt den Leser unmittelbar an die praktische Durchführung von AAS- und AES-Analysen heran, die in Kap. 6 näher beschrieben wird. Dabei wird besonders ausführlich auf die Monochromator-Konstruktion und die Untergrundkompensation eingegangen.

Jedes Kapitel schließt mit einer speziellen Literaturübersicht ab.

Das Buch ist — nicht zuletzt durch die zahlreichen Tabellen und Abbildungen — sowohl für Studenten der analytischen, anorganischen und klinischen Chemie, der Physik und Geologie, als auch für den Praktiker auf diesen Gebieten zur Einarbeitung in die Methode der AAS bzw. AES sowie zum Nachschlagen sehr zu empfehlen. Dazu tragen vor allem auch die Anhänge bei, in denen u. a.

Tabellen zu Nachweisgrenzen und Hinweise auf spezielle Anwendungen zu finden sind. Nützlich ist in diesem Sinne ebenfalls das Sachverzeichnis am Ende des Buches.

G. Hanschmann

Die ersten drei Minuten
Der Ursprung des Universums

S. Weinberg; übersetzt aus dem Amerikanischen von F. Griese. 5. Aufl. München — Zürich: R. Piper & Co. 1983. 267 S., 20 Abb., einige Tab.; gebunden DM 19,80

Rotverschiebung, *Doppler*effekt und *Hubbles* Gesetz von der einfachen Proportionalität zwischen der Fluchtgeschwindigkeit und der Entfernung von beobachteten Galaxien eines expandierenden Weltalls sind geläufige Erkenntnisse unseres Jahrhunderts geworden — weit bekannt und viel diskutiert. Und trotzdem gab es bis Mitte unseres Jahrhunderts zwei miteinander rivalisierende Erklärungen für *Hubbles* Gesetz: die big-band-Theorie und die steady-state-Theorie. Die Theorie vom stationären Universum forderte ein zu allen Zeiten gleiches Aussehen des Universums und bedurfte damit nicht der Einführung eines Startpunktes der kosmischen Entwicklung, wie ihn die Urknall-Theorie annehmen muß. Seit dem Nachweis der 2,7-K-Strahlung aus dem Weltall durch *A. Penzias* und *R. Wilson* hat sich diese Kontroverse, die sich bei manchen Untersuchungen und Überlegungen fördernd auswirkte, grundlegend geändert. Die in einem hohen Maß isotrope Hintergrundstrahlung deutet nicht nur auf ihren kosmischen Ursprung hin, sondern auch auf ein früheres Universum hoher (höchster) Dichte und Temperatur — ein untrüglicher Hinweis auf einen Urknall, der am Anfang aller Welten stand.

Der Autor, bekannt als Elementarteilchenphysiker, dem im Rahmen der *Weinberg-Salam*-Theorie die erfolgreiche Vereinigung der elektromagnetischen mit der schwachen Wechselwirkung gelang, unterzieht sich einer packenden Berichterstattung der Geschehnisse in den „ersten drei Minuten" nach erfolgtem Urknall. Das Universum ist eine „undefinierbare Suppe von Materie und Strahlung, jedes der darin enthaltenen Teilchen stößt sehr häufig mit den anderen Teilchen zusammen", und das nahezu noch vollkommene thermodynamische Gleichgewicht zeigt die unvorstellbare Temperatur von 10^{11} K (bei künftigen Neuauflagen sollte man die Gradangaben bei Kelvin-Temperaturen entsprechend den SI-Empfehlungen weglassen). Von diesem Zustand an wechseln in rascher Folge die Bilder: Bei der durch Expansion eintretenden Abkühlung und nach Unterschreitung bestimmter Schwellentemperaturen entstehen die uns bekannten Elementarteilchen.

Natürlich macht uns *Steven Weinberg* als Kenner der Elementarteilchenphysik auf noch vorhandene Unzulänglichkeiten dieses „Filmes der kosmischen Genese" aufmerksam. Da wäre der noch nicht beobachtbare kosmische Neutrino- und Antineutrino-Hintergrund von 2 K, was zu einer weiteren Stütze dieser Theorie führen würde — es sei ein wirklich quälendes Problem, meint der Autor: „... auf jedes Kernteilchen entfallen rund 1000 Millionen Neutrinos und Antineutrinos, und dabei weiß niemand, wie man sie aufspüren soll!" Zum anderen wird der Deuterium-Engpaß bei 10^9 K berührt, der bei genaueren Messungen des D/H-Verhältnisses im Weltraum eine Aussage über das anfängliche Verhältnis von Photonen zu Kernteilchen erlauben würde. Hier wirken sich die großen Isotopieeffekte des Wasserstoffs bei der Kondensation (beispielsweise in Form von H_2O, CH_4, NH_3 u. a. bei den großen Planeten) und bei Austauschreaktionen (beispielsweise zwischen Wasserstoff und Blausäure im Orion-Nebel) nachteilig aus und verfälschen in nicht leicht interpretierbarer Weise die wahren D/H-Verhältnisse.

Steven Weinberg ist zum Schreiben dieses Buches, das aus einem Vortrag entstand, den er im November 1973 an der Harvard-Universität hielt, überredet worden.

Wir dürfen seinen Freunden danken, daß sie ihn zum Schreiben dieses aufregenden Kapitels der Entstehung unseres Universums veranlaßten und es so einem weiten Interessentenkreis zugänglich gemacht haben. Es ist damit nicht nur eine moderne, mit geballter Sachkenntnis verfaßte Kosmogenie entstanden, sondern auch ein schöner Beweis dafür, wie eng fernste Welten der Astrophysiker mit den kleinsten Dimensionen der Elementarphysiker zusammengehören — eine zwar erwünschte Betrachtungsweise der Physik, die aber nicht jedem möglich ist. *Steven Weinberg* ist sie gelungen; er hat Hochinteressantes faszinierend dargestellt!

H. Hübner

Vom Sein zum Werden
Zeit und Komplexität in den Naturwissenschaften

I. Prigogine. München/Zürich: R. Piper & Co. 1979, 261 S., 66 Abb.; kart. DM 48,—

Das vorliegende Buch „Vom Sein zum Werden" von Nobelpreisträger *Ilya Prigogine* ist der Darstellung der Rolle der Zeit gewidmet und untersucht Möglichkeiten der Beschreibung zeitlich veränderlicher Prozesse im Rahmen einer allgemeinen physikalischen Dynamik und geht damit über die Frage hinaus, wie man mit zeitlich reversiblen Gleichungen, die gegen eine Zeitumkehr ($t \to -t$) invariant sind, zeitlich irreversible Prozesse beschreiben kann.

Sieht man vom Vorwort (in dem die Problem- und Zielstellung des Buches — die Irreversibilität nicht als eine Zusatznäherung in die Gesetze der Dynamik einzuführen, sondern die Dynamik in einen umfassenden Formalismus einzubetten — dargelegt wird) und vom Kapitel I ab (in dem einige grundlegende Aussagen zur Physik der Zeit in Zusammenhang mit dem zweiten Hauptsatz der Thermodynamik und der molekularen Beschreibung irreversibler Prozesse gemacht werden), läßt sich die Monographie in drei Teile gliedern.

Der erste Teil (Kap. II, III) stellt im Rahmen der „Physik des Seins" die Grundzüge der klassischen Dynamik und der Quantenmechanik zusammen, wobei der Autor insbesondere auch Grenzen und Entwicklung dieser Gebiete deutlich macht.

Der zweite Teil (Kap. IV—VI) bezieht die Zeit in die Betrachtung mit ein und untersucht im Rahmen der „Physik des Werdens" die Thermodynamik, die Selbstorganisation sowie Nichtgleichgewichts-Schwankungen und Ordnung aus dem Chaos. Hierbei werden besonders gleichgewichtsferne Vorgänge wie z.B. chemische Reaktionen dargestellt, Phänomene wie Stabilität, Verzweigungen oder Katastrophen betrachtet sowie Strukturen in Physik, Chemie, Biologie und Ökologie untersucht und ihre Behandlung vereinheitlicht.

Im dritten Teil (Kap. VII, VIII), der die mikroskopischen Eigenschaften (Teil 1) mit den makroskopischen Erscheinungen (Teil 2) koppelt, wird folgerichtig der „Brückenschlag vom Sein zum Werden" vorgenommen. Beginnend mit der klassischen kinetischen Theorie *Ludwig Boltzmanns* bis hin zu modernen mikroskopischen Theorien irreversibler Prozesse, die z. T. mit dem Namen des Autors verbunden sind, lernt der Leser eine große Palette von

Verfahren kennen, deren mathematisch-technische Einzelheiten im Kap. VIII bzw. im Anhang dargestellt werden.

Ein letztes Kapitel (IX) — die „Gesetze des Wandels" — gibt einen zusammenfassenden Ausblick.

Der Autor, der bewußt ein mittleres Niveau wählt, spricht einen großen Kreis allgemein naturwissenschaftlich interessierter Leser an, die über gewisse Kenntnisse der wichtigsten Methoden in Chemie und Physik verfügen. 139 Zitate verweisen auf weiterführende Literatur.

Dem Leser wird nicht nur ein faszinierender Überblick über die Problematik der Zeit in den Naturwissenschaften gegeben, er kann auch die Entwicklung der Methoden zur Beschreibung irreversibler Prozesse — die der Autor selbst maßgeblich mit gestaltet hat — miterleben.

Die gute Übersetzung des Originalwerkes "From Being to Becoming — Time and Complexity in Physical Sciences" von *Friedrich Griese* rundet den guten Gesamteindruck ab.

R. Haberlandt

Dialog mit der Natur
Neue Wege naturwissenschaftlichen Denkens

I. Prigogine u. I. Stengers: übersetzt aus dem Englischen von F. Griese. München—Zürich: R. Piper & Co. 1981. 314 S., 26 Zeichnungen; gebunden DM 39,80

Wir denken uns im Unterschied vom Raum, dem Nebeneinander, die Zeit als ein fortlaufendes Nacheinander, das in der klassischen Physik durch eine Gerade veranschaulicht werden kann, die nach rückwärts ins Unendliche, in die Vergangenheit, und ebenso nach vorwärts, in die Zukunft, verlaufen kann. Man sagt, *Ilya Prigogine* hätte durch seine Arbeiten — die 1977 mit dem Nobelpreis für Chemie ausgezeichnet wurden — die Zeit, den Zeitbegriff, „neu" wiederentdeckt. Dies ist zwar eine vereinfachende Formulierung, trifft aber insofern den Kern, als die Zeit mit der Qualität der Gerichtetheit, der Evolution und der Irreversibilität ergänzt wurde, — Begriffe, die gerade im letzten Jahrzehnt zu sorgfältigen und tiefgründigen Überlegungen Anlaß gegeben haben. Molekularbiologie, Elementarteilchenphysik, oszillierende Reaktionen und irreversible Thermodynamik begegnen sich in einer neuen, komplexen Art und können durch die genaue Analyse des Zeitbegriffes einheitlich verstanden werden.

Das Buch über das Gespräch mit der Natur gliedert sich in drei große Abschnitte: die Klassische Physik *Newtons* und *Laplaces* mit den Bahnkurven oder Trajektorien als zentralem Begriff, in welchem die Zeit als ungerichteter Parameter auftritt, den Abschnitt über die Wärmelehre des 19. Jahrhunderts (die Zeit der Entdeckungen der Erhaltungssätze) und schließlich den Hauptabschnitt: die Zeit seit der Entdeckung, Interpretation und Weiterführung des Begriffs der Entropie (die griechische Schreibweise auf S. 126 sollte ἐντϱοπη heißen). Eine scharfe zeitliche Trennung ist natürlich nicht möglich, hat doch selbst *Newton* drei unterschiedliche Konzeptionen des Kraftbegriffes gehabt, die er in einem längeren Prozeß vereinheitlicht hat. Und die Wärmetransportgleichung *Fouriers* ist auch lange vor der Herausbildung des Entropiebegriffes, aufgestellt worden. Was die Gerichtetheit der Zeit betrifft, so ist uns zwar sehr bald klar, ob ein vor uns ablaufender Film in der „richtigen Richtung" läuft, aber eine exakte Formulierung konnte erst durch die Einführung der Entropieerzeugung, durch irreversible Prozesse erfolgen. Die Abkühlung einer Tasse Kaffee, das Abbrennen einer Kerze und unser Altern sind solche zeitlich gerichteten Vorgänge, leicht erkennbar und unserer Erfahrungswelt entnommen. Solche zeitlich sich verändernden Systeme, dynamischen Systeme, können naturgemäß sehr komplexer Art sein. Die Dynamik eines Ökosystems, einer Landschaft, einer Stadt oder eines Industriegebietes deuten die Irreversibilitäten, die zeitlichen Entwicklungen und Veränderungen, an.

Die revolutionierenden Änderungen in den Ansichten der Wissenschaftler über die Welt, über Physik, Chemie oder Biologie beruhen häufig nicht so sehr auf neu entdeckten Tatsachen als vielmehr auf neuen Arten und Weisen, bekannte Tatsachen zu sehen, zu verstehen und von einem höheren Ordnungsprinzip aus zu betrachten. In diesem Sinne lassen uns die Autoren dieses überzeugend verfaßten Buches an einer neuen Betrachtung wohlbekannter Phänomene teilhaben. Es ermöglicht Forschern auch aus Nachbarwissenschaften, sich ein Bild über die wichtigsten Ergebnisse der Brüsseler Schule zu machen, die weithin als anerkannt gelten können.

H. Hübner

Atom- und Kernphysik

H.-U. Harten, H. Nägerl, H.-D. Schulte. Freiburg—Basel—Wien: Herder 1983. 128 S., div. Abb. (Reihe „studio visuell"); geb. DM 29,80

Kernkraftwerke und radioaktive Isotope haben in der Volkswirtschaft vieler Länder seit langem Bedeutung erlangt und bringen dort erheblichen Nutzen. Diese Begriffe sind ebenso wie der der Kernwaffe Millionen Menschen geläufig. Beide Richtungen, die friedliche Anwendung der Kernenergie und die Möglichkeit ihres kriegerischen Einsatzes, gehen naturgemäß auf die gleiche Basis, die Atom- und Kernphysik, zurück. Die Grundlagen dieses Wissenszweiges einem möglichst breiten Kreis näher zu bringen, ist Ziel des kleinen Buches „Atom- und Kernphysik". Dabei geht es an mehreren Stellen einen Schritt weiter als andere, unter einem ähnlichen Titel und mit den gleichen Zielstellungen erschienenen Bücher, indem z.B. in die Abhandlung der Kernenergiegewinnung auch die Probleme der Sicherheit von Kernkraftwerken und der Entsorgung einbezogen sind und im Abschnitt über Strahlenwirkungen ausführlicher auf die Dosimetrie und die natürliche und künstliche Strahlenbelastung eingegangen wird. Dagegen ist das zweite große Gebiet der Kerntechnik, die Anwendung radioaktiver Isotope, fast unerwähnt geblieben.

Das Buch ist in die vier Kapitel „Atome", „Atomhülle", „Atomkern" und „Elementarteilchen" untergliedert. Im dritten Kapitel sind dann solche Abschnitte wie z.B. „Messung radioaktiver Strahlung", „Kerntechnik", „Strahlenwirkungen" versteckt, eine vielleicht nicht ganz glückliche Lösung. Andererseits zeichnet sich das Buch durch eine einfache und klare Sprache und eine sehr instruktive Darstellungsweise in Schrift und Bild aus. An vielen Stellen erleichtern ergänzende Randerläuterungen sowie vergleichende Betrachtungen und Darstellungen auch dem weniger vorgebildeten Leser die Aufnahme und das Verstehen des in sehr kurzer Form wiedergegebenen Stoffes. Als potentieller Leserkreis dieses Buches sollten alle nicht unmittelbar auf dem Gebiet der Kernphysik und Kerntechnik ausgebildeten und tätigen Interessenten gesehen werden.

J. Holzhey

Elektronik, Lehr- und Arbeitsbuch

D. Zastrow. Braunschweig/Wiesbaden: Friedr. Vieweg & Sohn, 1983. 372 S., 463 Abb., 93 Lehrbeispiele, 162 Übungen; geb. DM 48,–

Es gibt heute nahezu keinen Zweig der Volkswirtschaft mehr, in den die Elektronik nicht Eingang gefunden hätte, und die Entwicklung verschiedener Bereiche, zu denen unbedingt die gesamte Kerntechnik gehört, ist in starkem Maße sogar direkt abhängig vom Stand und Fortschritt der Elektronik. Daraus ergibt sich ein breiter Interessentenkreis, bei dem der Wunsch nach einer zusammengefaßten, jeweils aktuellen und vor allem anwendungsbezogenen Darstellung dieses Fachgebietes vorliegt.

Das Lehr- und Arbeitsbuch „Elektronik" kommt diesem Wunsch weitgehend entgegen. Es vermittelt die Grundlagen der Industrieelektronik mit den Schwerpunkten Analogelektronik, Leistungselektronik und digitale Steuerungstechnik.

Die einzelnen Kapitel beginnen jeweils mit einer Aufzählung der zum Gebrauch des Buches erforderlichen Vorkenntnisse, um bei unter Umständen vorhandenen Wissenslücken diese durch gezieltes Nachschlagen relativ leicht schließen zu können. Es folgen eine kurze Stoffdarstellung sowie Beispiele und Übungen — letztere nehmen etwa 40% des Buchumfanges ein — und deren ausführliche Lösungen im Anhang. Besonders erwähnenswert ist — ebenso wie die spezielle Einleitung der Kapitel — das diese abschließende Memory, in dem das nach dem Studium des Buches oder einzelner Abschnitte erwartete Grundwissen enthalten ist.

Zum Verständnis des sehr übersichtlich dargebotenen Stoffes sind elektronische Vorkenntnisse nicht erforderlich, während ein solides Grundwissen in der Elektrotechnik vorhanden sein sollte.

Die vorliegende Ausgabe ist besonders als Lehrbuch, aber auch als Nachschlagewerk sehr zu empfehlen.

J. Holzhey

Proceedings of the Fifth Tihany Symposium on Radiation Chemistry

Hrsg.: J. Dobó, P. Hedvig u. R. Schiller, 2 Bände. Budapest: Akadémiai Kiadó 1983. 1142 S., 432 Abb., 132 Tab.; 1251 Lit.; gebunden Ft 990,–; DM 165,–

Das fünfte Tihany-Symposium fand im September 1982 statt, und bereits Mitte 1983 erschienen die umfangreichen Materialien des Symposiums. Neben den Vorträgen sind die Diskussionsbeiträge mit wiedergegeben. Die Einzelarbeiten wurden im Drucksatz nicht vereinheitlicht, die Herausgeber nahmen jedoch vorsichtig korrigierende Eingriffe in die Texte vor. Die Qualität der entstandenen Gesamtpublikation kann als sehr gut bezeichnet werden.

Die Symposiumsbände enthalten fünf Hauptabschnitte. Der erste Abschnitt enthält die Beiträge zur allgemeinen und theoretischen Strahlenchemie. Es wird über strahlenphysikalische Vorstadien und Begleiterscheinungen der Strahlenabsorption berichtet. Die Fragen der räumlichen Verteilung und Beweglichkeit der primären chemischen Produkte, sowie die Probleme der strahlenchemischen Kinetik und Mechanismen werden gemäß der Entwicklung der letzten Jahre an Beispielen oder in Übersichten dargelegt. Daneben wird jedoch eine Erweiterung des traditionellen Arbeitsgebietes angeregt durch Beiträge über Mühnen- und Positronenwechselwirkungen mit Materie. Auch die Reaktionen von implantiertem Tritium werden wieder in das Blickfeld gerückt.

Der zweite Abschnitt ist den wäßrigen und anorganischen Systemen gewidmet. Neben den bekannten Grundthemen über die chemischen Wirkungen der Primärprodukte der Wasserradioalyse auf gelöste Substanzen werden Fragen aufgegriffen, die aus dem Gebiet der Photochemie und Elektrochemie zu stammen scheinen, da es sich um Substrate handelt, deren Redox- oder Energieübertragungsvermögen in diesen Gebieten eine wichtige Rolle spielt (z.B. kondensierte aromatische KWS, Farbstoffe, Lumichrom, Metallophthalocyanine). Deutlich angesprochen ist das von *S. Solar*, *W. Solar* und *N. Getoff* (Wien), deren Thematik mit der chemischen Nutzung des Sonnenlichtes verbunden ist. Das neu erwachte Interesse an anorganischen Kolloidlösungen und gröber dispersen Metallen im Kontakt mit wäßrigen Lösungen wird durch entsprechende Symposiumsbeiträge deutlich sichtbar. Einzelerkundungen über strahleninduzierte Reaktionen von Lösungen einiger Nebengruppenelementsalze sind ebenso wie Grundlagenuntersuchungen zu präparativen oder technischen Prozessen nur wenig vorhanden.

Im dritten Abschnitt ist das umfangreiche Material über organische Systeme zusammengestellt. Es erweist sich, daß die Pulsradiolysetechnik genügend vervollkommnet wurde, um die Vielzahl der Reaktionswege in einem organischen System ordnend und aufklärend zu erfassen. Neben Radikalreaktionen werden zunehmend auch die Ionenreaktionen aufgeklärt; bedeutende Fortschritte bezüglich kationischer Prozesse und Zwischenprodukte werden dokumentiert. In diesem Abschnitt sind Praxisbezüge erkennbar, z.B. Beiträge zur Dosimetrie und zum Strahlenschutz, aber auch hinsichtlich präparativer (strahlenkatalytischer) oder technischer (Ethanchlorierung) Details.

Der vierte Abschnitt des Symposium-Materials enthält die Arbeiten zum Hauptanwendungsgebiet der Strahlenchemie, den Polymerisations- und Polymerenmodifikationsprozessen. 40 Beiträge geben einen recht umfassenden Überblick über den Stand des Gebietes. Die Propfpolymerisationen, die Erzeugung von Kompositmaterialien und die Untersuchungen der physikalischen und chemischen Eigenschaftsänderungen nach der Bestrahlung von Polymeren stehen im Vordergrund. Neue Trends in der Reaktionsführung oder Stoffauswahl sind nicht zu erkennen, vielmehr werden relevante Parameter an Einzelsystemen offensichtlich im Hinblick auf strahlentechnologische Entwicklungen studiert.

Der Schlußabschnitt ist strahlenbiologischen Problemen gewidmet. Einige Arbeiten beschäftigen sich mit niedermolekularen biologisch wichtigen Bausteinen und ihrem Radiolyseverhalten. Auffällig ist die Tendenz, nun auch hochmolekulare biochemisch wichtige Substanzen und Systeme bezüglich der Strahlenwirkungsereignisse genauer zu untersuchen (z.B. Lecithin-Liposomen, DNS, Stärke, Kollagen). Die strahlenchemischen Methoden scheinen solchen Aufgaben jetzt gewachsen zu sein.

Die insgesamt 145 Arbeiten geben einen guten Überblick über den Stand der Strahlenchemie 1982 in den sozialistischen und den wichtigsten übrigen europäischen Ländern. Die Bibliotheken sollten dieses umfangreiche Quellenmaterial in ihren Bestand aufnehmen, und alle an der Strahlenforschung beteiligten Wissenschaftler und Techniker sollten diese Quellen wie die der früheren Tihany-Symposien konsultieren.

U. Prösch

ISOTOPENPRAXIS

Volume 20 | 1984 | Number 8

Contents

Inhalt

Isotopenpraxis is indexed in Current Contents/PC & ES, Science Citation Index, and INIS Atomindex

FSC
www.fsc.org
MIX
Papier | Fördert gute Waldnutzung
FSC® C083411

Zeitfracht Medien GmbH
Ferdinand-Jühlke-Straße 7
99095 Erfurt, Deutschland
produktsicherheit@kolibri360.de